Cultural Studies of Science Education

Volume 17

The series is unique in focusing on the publication of scholarly works that employ social and cultural perspectives as foundations for research and other scholarly activities in the three fields implied in its title: science education, education, and social studies of science.

The aim of the series is to establish bridges to related fields, such as those concerned with the social studies of science, public understanding of science, science/technology and human values, or science and literacy. Cultural Studies of Science Education, the book series explicitly aims at establishing such bridges and at building new communities at the interface of currently distinct discourses. In this way, the current almost exclusive focus on science education on school learning would be expanded becoming instead a focus on science education as a cultural, cross-age, cross-class, and cross-disciplinary phenomenon.

The book series is conceived as a parallel to the journal Cultural Studies of Science Education, opening up avenues for publishing works that do not fit into the limited amount of space and topics that can be covered within the same text. Book proposals for this series may be submitted to the Publishing Editor: Claudia Acuna E-mail: Claudia.Acuna@springer.com

More information about this series at http://www.springer.com/series/8286

Jesse Bazzul • Christina Siry
Editors

Critical Voices in Science Education Research

Narratives of Hope and Struggle

Springer

Editors
Jesse Bazzul
University of Regina
Regina, SK, Canada

Christina Siry
Institute of Applied Educational Sciences
The University of Luxembourg
Esch-sur-Alzette, Luxembourg

ISSN 1879-7229   ISSN 1879-7237 (electronic)
Cultural Studies of Science Education
ISBN 978-3-319-99989-0   ISBN 978-3-319-99990-6 (eBook)
https://doi.org/10.1007/978-3-319-99990-6

Library of Congress Control Number: 2018968490

This Springer imprint is published by the registered company Springer Nature Switzerland AG.
The registered company address is: Gewerbestrasse 11, 6330 Cham, Switzerland

Contents

Part I
Critical Reflections: Stories of Struggle & Hope

Chapter 1
Critical Voices in Science Education

Jesse Bazzul and Christina Siry

Many people need desperately to receive this message: 'I feel and think much as you do, care about many of the things you care about, although most people do not care about them. You are not alone'.
– Kurt Vonnegut Jr., Timequake (Vonnegut 1998)
I dream. Sometimes I think that's the only right thing to do.
– Haruki Murakami, Sputnik Sweetheart (Murakami 2002)

This book came into being because of an ever-present cacophony: the sounds of students, teachers, and researchers critically engaging the field of science education for the wellbeing of communities and justice for our shared planet. It is no easy task, as hardship, resistance, and confusion admittedly account for some of these sounds. There are no promises of return (of any kind), no roadmaps to tell where and how something might be changed in a field that could be so integral to the creation of environmentally and socially just futures.

This collection captures the diverse stories and journeys of science education scholars as they have come to do important critical work in the field. Work that can often be opposed, censored, or discouraged by institutions, social forces, and even people we have come to trust and learn so much from (and still do). What follows are narratives of struggle, sense-making, and hope generously shared by a diverse group of science teacher-educators and science education researchers. Collectively, they present snapshots of their various experiences, as well as insights into the challenges many justice-oriented science educators face as they work within education systems that keep systems of oppression and destruction, such as white supremacy

J. Bazzul (✉)
University of Regina, Regina, Saskatchewan, Canada
e-mail: Jesse.Bazzul@uregina.ca

C. Siry
Institute of Applied Educational Sciences, The University of Luxembourg, Esch-sur-Alzette, Luxembourg
e-mail: Christina.Siry@uni.lu

J. Bazzul, C. Siry (eds.), *Critical Voices in Science Education Research*, Cultural Studies of Science Education 17,
https://doi.org/10.1007/978-3-319-99990-6_1

and global neoliberal capitalism, locked in place. The goal of the book is simple, yet, we feel, expansive and of-the-heart. It is intended as a beacon of positivity, guidance, and faith to anyone taking a tumultuous ethical road in the field of science education. In solidarity – you are not alone!

1.1 Meaningful Connections and Dark Basements

Jesse: *The idea for a 'critical voices' book came at a time when doing justice work in science education seemed extremely daunting to me as a recent PhD graduate. I did not have a job, and prospects were uncertain at best. I began to doubt whether science education was a real discipline! Was the field simply a distraction from the necessary justice work needed to avoid immanent catastrophe – social, environmental, and spiritual? Should vital energies be exerted elsewhere?*

I was fortunate enough to be able to discuss these questions with Carol-Ann Burke when we were students at the University of Toronto (and later, long distance between Calgary/Toronto, Canada, and Massachusetts, USA (see Burke and Bazzul 2017*). Our conversations about science education led us to think that it might be helpful to assemble the voices of science education researchers, graduate students, and teacher educators dealing with similar dilemmas. Though Carol-Ann needed to step away from this project, I am indebted to those conversations. For me, they are an example of the meaningful connections educators seek to have with a community; connections that are absolutely necessary to being and thinking. Now, the collaborative work with Chris Siry, and the personal work shared by authors, have become another series of meaningful connection points. In fact, this whole book might be best thought of as a constellation of meaningful connection points that are simultaneously different, yet appeal to commonalities we might have in doing this work. To put it broadly, these points begin to form images and dreams of, 'how things might/should go'.*

One thing that is woven into everyday educational life are tiny glimpses of living-in-common and the desire to share as much as possible. For example, in my experience, the smaller side of religions teach me all about a radical kind of equality, yet often they do so by challenging the very structures that allow these ways of being to exist in the first place. The education building at the University of Toronto where I studied was brutal, but in and around such places people are creating something different and wonderful. I need people to show me strange things, and teach me love. As educators, we must continually create connection points, make them grow, and proliferate the power they bring.

Chris: *When you first asked me to join this project, Jesse, I was excited to be able to collaborate on creating a space for authors to critically reflect on their experiences and share these reflections; a space for giving voice to their stories. The resulting stories are intensely personal, and illustrate the diversity of ways in which* *we deal with struggles while continuing to seek opportunities to make and remake our worlds. These deeply personal stories create biographical reflections which capture "…a vast array of impulses, instincts, memories, and dreams – visualized, theorized and told as a story" (Dhunpath* 2000, *p. 546). In the diversity of stories shared in the chapters of this book we find the connection points you write about, which underscore the entanglements between our experienced realities. These layer, twist and wind together to give a lens into "how things are" through the narratives shared in the sections that follow, from which I emerge as a reader with my own reflections on "how things could/should be." A book creates a relationship, a relationship between the writer and the reader, and those readers who engage with the stories in this volume will hopefully emerge with new reflections and considerations about their own work. The narratives that follow each reflect individual challenges encountered by those working to "do good work" while collectively coming together to leave a reader with a sense of hope; hope that through the expression of these polysemic reflections on struggle and resistances we can come together to support each other in creating connection points to collectively find openings for transformations.*

1.2 The Trouble with "Doing the Work" and the Power of Collective Voice and Narrative

The chapters of this volume highlight a multitude of ways in which scholars in the field of science education are making sense of the journeys they have undertaken, and together highlight some of the troubles experienced by those "doing the work". The authors express the power that can come from turning to diverse critical theories to make sense of experiences and to use this sense-making to work towards transformations. "We humans have a deep relationship with our past experiences, as 'each occurrence is charged with echoes and reminiscences of what has gone before, where each event is a reminder of other things' (Dewey 1920, p. 1)" (Goodson and Gill 2014, p. 224). One of the things this volume achieves is a tangible, and somewhat bulky, move toward privileging narrative, voice, and creativity. We do not tease out narrative from storytelling or narrative inquiry. We have also not refined our use/definition/description of narrative, nor have we requested our contributors to do so. In broader terms, however, and in a way that is relevant to 'science people', this volume erodes the separation of what philosopher Jean-Francois Lyotard (1984) deemed scientific knowledges and narrative knowledges. Many scientists, as well as social scientists, may dismiss the privileging of narratives and/or narrative knowledges and view them as the non-province of science and, by extension, science education. This may be because narratives and narrative knowledges have no (or few) methods of internal legitimation. However, following Lyotard, one vital point about narrative and scientific knowledge still remains relatively unconsidered: if science and science education want to justify their engagement in the world, this can only happen in the form of narratives and narrative knowledges – the stories we choose to tell and create. As educators we have an ethical role to engage the narratives that guide and constitute what and how science is done, who science is for, and for what end(s). The stakes and complexity of this task could not be higher.

Storying can be an endeavor that is oriented towards liberation and transformation (Goodson and Gill 2014); one in which we are reminded of the past as we work to construct the future. We facilitated the creation of this book, not as a stable or universal narrative on struggle or resistance, but rather, to underscore the value of storytelling and the power of reflecting on the stories we are told. In writing about life history research, Dhunpath (2000) coined the term "narradigm" to illustrate the ways in which "…our lives are intrinsically narrative in quality. We experience the world and re-present our experience narratively" (p. 545). The chapters that follow use a wide range of genres and authorship to re-construct, re-present and interpret experiences. Story, metalogue, poetry, art, and theory all come together to elaborate the distinct perspectives and stories of the authors. Taken as a whole, these chapters reveal a little of the complexities of being and becoming for critical science educators. They also elucidate some of the subjective meanings people ascribe to their experiences, along with how these shape individual and shared perspectives on the role of science education scholars. The narratives of resistance to dominant paradigms, drawing on theories of resistance and emancipation, prompt reflections on notions of what "science" is, and what it means to be critically oriented researchers of science education. It is our hope that this polyvocal/polysemic book illustrates the ways in which our historicity shapes how we teach, what we write, and how we conduct research; in short, 'who we are' as science education researchers.

References

Burke, L. E. C. A., & Bazzul, J. (2017). Locating a space of criticality as new scholars in science education. *Cultural Studies of Science Education, 12*(3), 565–579.

Dhunpath, R. (2000). Life history methodology: "narradigm" regained. *Qualitative Studies in Education, 13*(5), 543–551.

Dewey, J. (1920). *Reconstruction in Philosophy.* New York: Henry Holt and Company.

Goodson, I., & Gill, S. (2014). *Critical narrative as pedagogy*. New York: Bloomsbury Academic.

Lyotard, J. F. (1984). *The postmodern condition: A report on knowledge* (Vol. 10). Minneapolis: University of Minnesota Press.

Murakami, H. (2002). *Sputnik sweetheart*. New York: Vintage.

Vonnegut, K. (1998). *Timequake*. New York: Penguin.

Jesse Bazzul is Associate Professor of Science and Environmental education at the University of Regina. He feels that comprehensive attention needs to be given to the way typically depoliticized fields of study, such as science education, work to constitute political and ethical forms of life.

Christina Siry is an elementary science education researcher and teacher education, and currently Professor of learning and instruction at the University of Luxembourg. Her research is grounded in critical theoretical approaches and she has a particular interest in plurilingual students' interactions as they do science at the elementary and early childhood levels.

Chapter 2
Boundaries, Gatekeeping, and Oppression Within Science Education Research

Sarah Riggs Stapleton

In reflecting on the field of science education, Bazzul notes that, "examining our role as critical educators in a conservative field involves negotiating crucial tensions between complicity and obligation, participation and refusal" (Burke and Bazzul 2017). I will be addressing these tensions within this entry. In trying to articulate my frustrations as a scholar still new to the field of science education, I realized that my concerns demanded a medium better suited to carrying emotion than academic prose. Through poetry, I hope to raise questions, surface issues, and highlight voices. With poetry, I also invite others in to create space for their own concerns, tensions, and wonderings about the field. Throughout, I raise questions about how we engage in social justice work in light of the gatekeeping represented in the poem. I draw primarily upon experiences I had as a graduate student, when the stifling pressure to conform nearly ended my academic career before it had begun.

2.1 Boundaries, Gatekeeping, and Oppression Within Science Education Research

Why hold each other back?
        Alternate ~~IDEAS~~
                ~~VISIONS~~
                ~~DREAMS~~

S. R. Stapleton (✉)
University of Oregon, Eugene, OR, USA
e-mail: sstaplet@uoregon.edu

J. Bazzul, C. Siry (eds.), *Critical Voices in Science Education Research*, Cultural Studies of Science Education 17,
https://doi.org/10.1007/978-3-319-99990-6_2

Research [this].
Your questions aren't welcome.
That's not science education.
You won't get a job
doing *that* as a dissertation.
[*This*] is how we do social justice.

No R O O M for other voices
othered voices
nonconformists
dissonance
self-reflection
criticality

No R O O M for **you**.

Citations a popularity contest.
Victors = unexamined oppressors
of many
(especially graduate students
first generation
women
of color).

Bruised, the oppressed
d dignity, self-worth
l
i
u
must reb
wondering if they belong.

Is this just?
Conforming.
Complacency.
Complicity.

Meanwhile…
Researchers in insular conversations
not realizing
no one else cares.

Shouldn't scholarship
b k new ground?
r a
e a k new ground?
a
k new ground?

Neoliberalism
        framing STEM.
Not fought,
pointed out,
even…realized?
        [Sorry... too busy.]

Standards zealotry.
Can standards leave R O O M
        for the non-standard?

Environmental justice.
Climate change.
        On shoulders
        but not minds.

Egos
Grants
Selves
        Is *just* research possible?

Reference

Burke, L., & Bazzul, J. (2017). Locating a space of criticality as new scholars in science education. *Cultural Studies of Science Education, 12*(3), 565–579.

Sarah Riggs Stapleton is an Assistant Professor in Education Studies at the University of Oregon's College of Education. She earned her PhD in Curriculum, Instruction, and Teacher Education from Michigan State University. She also holds a master's degree from the Harvard Graduate School of Education. Sarah uses critical and participatory methodologies to investigate social and environmental inequities. She is particularly interested in food and agriculture as learning contexts for science and environmental education. Prior to her doctoral work, Sarah taught AP Environmental Science, chemistry, physical and general sciences at the middle and high school levels in public schools in California and the Gambia, West Africa.

Chapter 3
Quietism in the Face of Injustice: A Cultural Mennonite's Reflection on Pride and Shame in Science and Environmental Education

Hannah K. Miller

> *Go placidly amid the noise and haste, and remember what peace there may be in silence. As far as possible, without surrender, be on good terms with all persons.*
> –Max Ehrmann, *Desiderata*

This first line of Max Ehrmann's 1927 poem *Desiderata* is the first thing guests see as they enter my childhood home. Although not written by a Mennonite, the poem embraces the Mennonite commitment to pacifism, silence, and good cheer in the face of the world's problems. Though my culturally Mennonite childhood taught me to value quietism, I take issue with a pacifist approach to citizenship that makes no space for critical interrogation of injustice and deliberate action towards dismantling oppressive systems in education. As an emerging scholar in the field of science and environmental education, I see numerous injustices that demand resistance and action for change. This terrifies me. The following autoethnographic reflection examines a conflict between two selves: a researcher who wants to commit to critical methodologies, and a cultural Mennonite who views confrontation as a high sin.

3.1 Scholarship That Has Helped Me Frame This Story

To guide the construction of this account, I draw on Ellis et al.'s (2011) concept of understanding autoethnography as an analysis of the marriage between self and culture. The context in which this particular autoethnographic account piece is situated, however, adds additional theoretical layers. One of these is my commitment to exposing power and privilege in educational systems through critical, anti-racist,

H. K. Miller (✉)
Northern Vermont University, Johnson, VT, USA
e-mail: Hannah.Miller@NorthernVermont.edu

J. Bazzul, C. Siry (eds.), *Critical Voices in Science Education Research*, Cultural Studies of Science Education 17,
https://doi.org/10.1007/978-3-319-99990-6_3

feminist perspectives, which examine the "connection between schools and class interests, patriarchy, and race" (Weiler 1988, p. 2).

Another layer comes from my use of Critical Race Theory (Delgado and Stefancic 2012). Specifically, I use interest convergence as a frame to examine my own complicit participation in racist and heterosexist systemic oppression. Interest convergence is visible when members of a dominant group benefit from anti-oppressive actions or policies. Having the backing of the dominant group provides the structural power that justifies the action or policy actions. The primary beneficiaries of the actions are those in power, not those the actions purport to serve. In the context of race, for example, interest convergence is visible when "white elites tolerate racial diversity advances only if it benefits their own individual or group interests" (Hughes and Giles 2010, p. 47).

Finally, I agree with Carolyn Ellis (1999) that autoethnographers must commit to vulnerability. While white and otherwise privileged scholars fighting for justice in oppressive systems may invite vulnerability (e.g., speaking back to oppressive narratives from within education threatens their position of power and privilege within that system), I find that a more salient form of vulnerability emerges by examining my own implicit support of systemic racism and heterosexism in my work as a scholar. In other words, the vulnerability I hope to reveal in this chapter stems from a critical examination of my own participation in supporting oppressive systems, not a direct critical examination of the oppressive systems themselves.

3.2 Who I Am, Questions I Ask, Stories I Tell

This autoethnographic account discusses how my methodological perspectives as a researcher— as well as my ontological positionalities as white, American, cisgender, homosexual, able-bodied, English-speaking cultural Mennonite—complicated my research in an undergraduate environmental sustainability education program in a Mennonite college in the rural Midwestern US. Data sources include autoethnographic field notes and a written exchange between myself and my advisor about our conflicting perspectives on the value of Critical Race Theory (CRT) in education. The chapter is divided into three sections, each of which is oriented around a specific conflict that emerged during the research: (a) facing my Mennonite cultural heritage and fear of conflict, (b) the Mennonite rejection of the LGBTQ community and my complicit support of heterosexism, and (c) facing my complicit support of racism in science education. The first story is one I am proud of, because it demonstrates a successful example of how I was able to use my own privilege to fight injustice. The other two, however, expose my own complicit acceptance of oppression. These stories keep me up at night.

3.3 Conflict 1: Facing My Own Mennonite Cultural Heritage

I began my graduate career by returning to a historically Mennonite college as both an insider (having been raised by Mennonites) and as an outsider (having not adopted their faith and traditions due to philosophical differences). My grandparents were descendants of the Swiss-German Anabaptists who migrated to North America to escape persecution in Europe for their commitment to adult baptism. As a reminder of the suffering that accompanied this persecution, many churches and homes display copies of the ominous tome The Martyr's Mirror (Braght and Sohm 1987) which documented the bloody martyrdom of European Anabaptists. When these pacifists migrated to North America, they were worried the government might threaten their spiritual independence. My grandparents spoke Pennsylvania Dutch, a language which was preserved in North America as a result of the newly settled North American Mennonites' social agreement with various civic bodies that they could maintain religious and civic sovereignty through retreat to separatist, quietist, agrarian lifestyles (Driedger and Kraybill 1994).

Due to his Mennonite upbringing, my father was a Conscientious Objector to US involvement in the Vietnam War. Having successfully defended his case for conscientious objection, he was sent, with other Mennonites, to urban Atlanta to teach in underserved schools. During this time of service, he and my mother constructed a community with other Volunteer Service members, where their distance from elder Mennonites allowed the space to raise critical questions about the value of quietism (i.e., pacifist retreat from mainstream society) in a world that was desperate for activism, justice, and change. It was in the legacy of this community's critical examination of their own heritage that I was raised. Despite my contentment as a member of this community, when I graduated from college I sought distance from the constraints of my Mennonite heritage. Although I grew up relatively removed from the cult-like social regulations that marked my grandparents' childhoods, the pressure to "*go placidly amid the noise and haste*" stuck with me. I did not attend a Mennonite institution of higher education for my undergraduate studies, and upon graduation I moved to China where I stayed from 2001 to 2009 teaching and studying the Chinese language and culture.

During my time abroad, I started organizing small gatherings for the hidden Shanghai LGBTQ community. Over time, we built a social network for the underground queer community, and the year before I left we decided to organize a Pride event. After what turned into a terrifying face-off with the national government, my leadership team and I had organized what is now known as China's first successful gay pride event, which recently saw its 10th consecutive celebration (Jacobs 2009; Liu 2009; Comrades-in-arms 2009; Lim 2009; Hogg 2009; Weihua 2009). As the week of our Shanghai Pride festival was unfolding, we the organizers were certain that I was headed to prison. Police took business licenses away from our supporting vendors. Organizers were followed by plainclothes police during the events. I was advised by our most trusted legal advisers and press correspondents to delete the contacts on my phone and avoid communication with fellow organizers for fear of incriminating them. I was advised that my communication was being monitored. I notified my principal that police might arrive at the elementary school where I taught to arrest me. I was terrified.

Despite my fear, I did not back down. Confronting the Chinese government forced me to reconcile my yet unexamined internal conflict between my cultural values of pacifism and quietism with my thirst for challenging injustice. Backed with the privilege of white American status, this conflict was relatively safe to embrace. I felt comfortable taking on the role as the public face of the festival because I was confident that even if I spent a lifetime in jail, I would not be executed, and I would receive preferential treatment. As a young person, my path was not yet set, and fighting for justice from within Chinese prison felt like a worthy path.

Shanghai Pride is a story I am proud of. In this story, my fear of conflict was not a handicap. I was loud. I rejected the peace of silence. To our surprise, I was not arrested, and I returned to the U.S. to begin graduate school the day after the festival ended. The two stories that follow, which took place in the context of my research and scholarship, are stories I am not proud of. How do I reconcile a willingness to face the Chinese government before my career as a scholar (a source of pride), and then a willingness to comply with systemic heterosexism and racism in my scholarly work (a source of shame)? This question is what makes this autoethnography necessary.

On the final day of Shanghai Pride, finding myself not in prison, I returned to the United States to begin graduate school at Simons College (all names are pseudonyms), a historically Mennonite institution of higher education. My success fighting for justice in China compelled me to return to the place of my roots to reconcile the inner conflict between pacifist and activist. In my time at Simons College I was intrigued by similarities between the philosophies of the Simons community and my own identity. Some were more superficial expressions of consumerism (e.g., the value of organic, unprocessed food, and simple, unadorned clothing) while others were more deeply rooted in my ethos of citizenship (my rejection of war and the military; my disdain of the American flag and demonstrations of patriotism). To my surprise, my own identities were heavily influenced by the Mennonite socio-cultural ethos. I had not escaped. I was an insider. This raised a question: How could a social activist be an insider in a pacifist, quietist space?

Soon after asking this question, I found myself writing a dissertation proposal to study a Simons College environmental education program (Miller 2016). Because of my own struggles to reconcile my pacifist and activist selves, I thought that examining this same tension within an education program might help me reconcile this personal conflict. I wanted to examine how an educational program at a college that valued silence and pacifism could commit to the study of environmental justice and sustainability, fields which demand action. I wondered: How do pacifist Mennonites, when faced with an unjust and unsustainable world, respond?

3.4 Conflict 2: Facing My Own Passive Support of Systemic Heterosexism

The following story is something I am not proud of because it exposes ways I have used my pacifist values to avoid conflict and back away from fighting for justice. During the time of my dissertation research, Simons College was in the process of examining its official stance towards LGBTQ faculty. The Mennonite church was threatened with a schism over the potential acceptance of queer pastors (of their institutions of faith) and faculty (of their institutions of higher education). Situated in a culture that had mastered the art of conflict avoidance, they had few models for how to reconcile such an internal battle. Many historical conflicts within the church are best understood through written record, a medium through which people felt more comfortable challenging each other's ideas. This is an important point: historically, the Mennonite culture has not lacked conflict; it is the confrontation and resolution of these conflicts that they have struggled to embrace.

In what appeared to be a step toward action and confrontation, the college hosted a series of "listening sessions," during which people shared perspectives on whether or not the college should include LGBTQ faculty in its non-discrimination policy. During these listening sessions, queer faculty at the college (including my wife) were silent for fear of losing their jobs. The listening sessions, because of the college's hostile treatment of LGBTQ faculty (some had been verbally threatened, some had been fired, and some had quit, according to local stories that circulated among the LGBTQ community), excluded the LGBTQ community members' voices. The listening sessions were not designed to listen to everyone.

During this national discussion, a professor at an Anabaptist college and Mennonite historian spoke on a National Public Radio story (Reddy 2015) in response to LGBTQ issues facing the church, saying that "fast social change is dangerous for a community." This enraged me. Fast social change is only dangerous for those who benefit from systems of oppression and risk losing privilege. I wanted to call out the hypocrisy of the church and its leaders. I wanted to yell that their unwillingness to rock the boat was a privilege in itself, and that those who were silenced by the oppressive structures in the Mennonite church could not see still waters from any direction and did not share their complacency with oppression. I wanted to condemn their cowardice and violence towards the LGBTQ community.

With my work from Shanghai Pride in mind, I felt compelled to speak out at these sessions to expose the hypocrisy of a religion that prided itself on its moral virtues locally and abroad, yet perpetuated injustice within its own doors. However, I feared for my wife's job, and I feared for my own continued position as a guest on their campus where I was conducting my research. I slowly started to realize that I was no different from the Anabaptist professor who was unwilling to rock the boat for fear of uncomfortable consequences. I was silent.

This silence is something I regret daily (a source of shame). I regretted my presence on their campus. I regretted my membership in their community. I regretted my complicit support for heterosexism and discrimination. In reflecting on my passive support of heterosexism, this question emerged: Why was I willing to use my privilege as a white American abroad to fight for change during Shanghai Pride, yet unwilling to use my privilege as a white American Mennonite to face the church and Simons College? I was not silent in China, despite the risk of jail in a foreign country and the possibility of never seeing my family again. These were considerable consequences. Why did the relatively benign fear of a prolonged dissertation (needing to find another place to conduct research) and my wife's potential job loss (which she could have surely found again, considering her experience) suddenly become too great a consequence? As members of white middle-class America with financially stable families, why was losing a job and risking a prolonged dissertation enough of a risk to remain silent in the face of oppression?

I have considered many possible explanations for this. Did living in the geographical center of the North American Mennonite Community contribute to the incremental weakening of my commitment to social justice? Had becoming an insider in that community pose a threat to my identity as an activist? In China, I was clearly an outsider. Fear of being banished from a community in which I never truly belonged posed little threat to my identity. Standing up to the Mennonite church at this point in my career, however, meant loss of acceptance from my cultural and, perhaps, academic peers. This forced me to realize that my Mennonite cultural heritage was a more central part of my identity than I had previously thought, and the potential loss of my inside status in that community scared me. I hated myself for compromising my deeply held commitment to social justice (a source of shame). I was complicit in perpetuating systemic bigotry and heterosexism.

As I reflect on this silence, the concept of *interest convergence* from CRT offers some explanatory power for my inaction at Simons College. I was willing to take a stand against homophobia and oppression as long as my own privilege was not disrupted. I was willing to speak my truth behind closed doors or around bonfires with my lesbian peers, but publicly speaking out against the college no longer served my own interests. The selfish needs to advance my own scholarship and maintain financial stability outweighed my commitment to fighting injustice. This is a source of shame.

3.5 Conflict 3: Facing My Own Support of Systemic Racism in Science Education

This final story is another story of silence (a source of shame). This is a story of my own passive support of systemic racism in science education. Because of my affinity to feminist, queer, and critical perspectives in science education, I think of myself as someone committed to the dismantling of the white supremacist, heterosexist, and anti-feminist systems within the field. This story, however, makes me question my commitment to this work. This story is not over, however, and through this reflection I hope to forge a path forward towards reconciliation.

With the fight within the heterosexist Mennonite Church USA operating in the background, I managed to complete my dissertation data collection successfully and was excited to dig into my analysis and discussion. Although I began with a theoretical framework of agency/structure dialectic, I found in the analysis that CRT had more explanatory power for certain components of my dissertation. When I brought these ideas to my advisor, he questioned the value of CRT. The conversation that followed (which I include below) made me question my commitment to dismantling oppressive systems within science education and also helped me more clearly identify work that is left to be done. First, I offer a brief overview my dissertation findings and interpretation to provide context and history.

3.5.1 Background

My dissertation was a case study of an environmental sustainability educational program at Simons College. One of my student participants, Franco, (pseudonym), experienced numerous encounters with racism and xenophobia during the program. Although Franco (a non-white, non-American, cisgender heterosexual male) recounted numerous stories of racism during his time in the US, this section contains just two of these that took place during the program.

The first was that when one of the faculty members, Alfred, (pseudonym), admonished Franco for not tailoring a course presentation to an American audience. Alfred criticized Franco's limited vocabulary and language (Alfred couldn't understand some of the presentation), as well as what Alfred described as Franco's failure to tailor his presentation to an American audience. My field notes from these classes were tense and angry. From my interpretation, this was a prime example of whiteness as property (Ladson-Billings and Tate 1995), in that the white audience and perspective demanded that other narratives conform to white ways of knowing.

The second happened when Franco went to a neighboring community to conduct a series of interviews about sustainability. It was Franco's task to make observations, interview townspeople about their attitudes towards sustainability, and construct a proposal (based on his interview results) for the townspeople about how to address their sustainability concerns. Although this was not mentioned or discussed prior to the assignment, as could be expected, Franco's race, ethnicity, and language profoundly impacted his experience during this assignment. During his interviews, Franco was told by the local townspeople that Marietta was a "white settlement." On his way out of town, a sheriff pulled Franco over on his bike, searched Franco's bags, demanded to know what he was doing there, required him to produce identification (which he didn't have), and informed him that there had been complaints of a suspicious person in the area.

Franco told the faculty and his classmates about this experience, and the faculty (all of whom were white) were concerned, but expressed their concern in ways that raised red flags. Using CRT as an interpretive lens, I identified a few characteristics in their responses that were important for me to include in my analysis of patterns identified in CRT scholarship (Solorzano et al. 2001). For one, they took a "color-blind" approach (DeCuir and Dixson 2004), suggesting that Franco's mistreatment was due to his outsider status, not his race or ethnicity. They further suggested that this experience was typical of the generic "human struggle" to accept outsiders, and that they as white men had faced similar struggles in their life to be accepted (Baber 2015). Second, they denied the permanent and systemic nature of racism (Ladson-Billings 1998) by suggesting that the event was a result of a few ignorant individuals in an otherwise racially harmonious community (Miller 2018).

When I interviewed faculty members about this incident, I did not respond to their ideas immediately. I felt uncomfortable and unprepared to evaluate their responses in the moment. I was scared to say: "Franco's story is about race. Why are you removing race from this story?" Instead, I used the analysis in the dissertation to express these ideas. To rectify this lack of immediate response, I organized a series of workshops with the faculty to process my results. Together, we found a language for talking about systemic racism within the program, and region, sustainability education, and the surrounding community, and the field of environmental education. I felt we were doing good work (a source of pride).

3.5.2 Exchange

When I shared this discussion with my advisor, his response angered me. I found his response typical of how white liberal academics address racism in education, and also typical of patterns CRT scholars have identified as a barriers to achieving racial justice in education. When I read his initial response I thought: "how is it that I've been working with this person for five years and this perspective on race and justice has never surfaced?" I was ashamed that I had not pushed for these conversations earlier in my career. I began to realize that my willingness to let these issues of justice remain hidden during my research and work made me complicit in the perpetuation of systemic racism in science education (a source of shame). I felt the same rush of anxiety return that I felt when I had stayed silent about the Simons College LGBTQ faculty policy at Simons College.

I include this exchange (Figs. 3.1 and 3.2) to make conversations that take place behind doors in science education visible. This exchange is not a reflection on "a white student's fight against her advisor's racist comments," but a reflection on how my position as a white researcher made it possible for me to benefit from the power and status of my advisor's position in the field by implicitly accepting the system (of which these ideas were just one symptom). As a white doctoral student, I benefited immensely from my advisor's experience and status in the field. I am implicated in this system of racism, power, and oppression. The presentation of this exchange is an attempt to bring my own role in this system to the surface for deconstruction and analysis. With his permission, I share (below) our exchange in which we express opposing ideas about the value of CRT in my dissertation (specifically) and in science education (in general).

"Hi Hannah--
I'm not sure how I can help with the narrative that you are developing here. On the one hand, I recognize CRT as a legitimate and respected form of scholarship. On the other hand, I can't agree with the balance between the perspectives and insights that it privileges and the perspectives and insights that it criticizes. There's no way to make this critique without positioning myself as an old white guy who doesn't get it, but I'll make some comments anyway.

CRT as American exceptionalism. CRT feels to me like a version of left-wing American exceptionalism that has some of the same problems as the right-wing version. Well-educated people of color in the US are among the world's most privileged people. It's important to note that they are less privileged than well-educated white Americans, but I see other forms of injustice as deserving higher priority. I get particularly uncomfortable when CRT is used to portray more privileged people, like Franco, as being oppressed by less privileged people, like the citizens of Marietta. So I believe that there are people whose opportunities in life are denied or constrained by racism and xenophobia, but I have trouble seeing Franco as one of those people."

The role of education in an unjust world. Students come to educational programs with talents and opportunities that are unfairly distributed. And the programs must prepare them to succeed in a world whose institutions are also unfair. Figuring out what the appropriate role and stance is for educational programs is in this world is complicated, but I don't think it's clear that they are more effective if they are islands of more perfect justice. So I see a narrative here in which the [Sustainability Semester] is doing a pretty good job for Franco:

With respect to his narrated dialectic, **his experiences are leading him toward the correct conclusion that the rural American Midwest is not where he can be most effective.**

With respect to his embodied dialectic, **he is learning lessons and acquiring skills that are essential for his later success in South Africa and other contexts. For example, he didn't like Alfred's critique of his presentation, but I have to wonder whether that contributed to the way he prepared for successful presentations later. Similarly, learning to be cautious in parts of the world where the local people are racist, xenophobic, and carry guns may be a useful lesson for someone who aspires to a future that brings him into contact with many different cultures.**"

So I recognize that you know Franco better than I do, and I can't claim that my interpretations are any better than yours. (It's a tribute to your storytelling that you have presented data that are rich enough to support the construction of alternate narratives.) But these are the differences I see between the narrative I would construct and the one you are developing in the Franco excerpt. I encourage you to continue, but I don't think I'm the right person to help you with this particular analysis. Andy (23 May 2016, personal email).

Fig. 3.1 My advisor's response to use of CRT as an interpretive lens in my dissertation

I cowered when I read my advisor's email (Fig. 3.1). I was embarrassed that he felt comfortable sharing these ideas with me. I was embarrassed that I was a part of this exchange. I thought: "Would he ever share these ideas with a person of color?" From this exchange emerged another source of shame: my initial decision was to *not* respond. After all, he said that he was "not the person to help [me] with this particular analysis." That gave me a clear path out of this battle. For the following week, each morning when I began my work, I rehearsed what I might like to say in response to this email were we to discuss it. I practiced in front of the mirror. I drafted emails. Each rehearsal ended in tears. I felt it was unprofessional to respond with raw emotion. I was afraid to embarrass myself. I was afraid to tarnish my reputation. I was afraid to disappoint my advisor.

I am not proud of this moment in my graduate career (a source of shame). This was clear evidence to me that although I claimed to support the scholarship of feminist, anti-racist, critical scholars in the field of science education, I realized I might be too fragile for this work. The same question I asked myself during my silence with the LGBTQ faculty policy fight at Simons College resurfaced: Why was I so willing to go face to face with the Chinese government during Shanghai Pride, yet

First, I disagree with the notion that Franco is "more privileged" than white Marietta citizens. This hierarchy of oppression, or the "oppression Olympics" (Yuval-Davis, 2012), discredits accounts of oppression from a speaker in a world that contains individuals who are "more oppressed" than the speaker. One outcome of constructing such a hierarchy of oppression is that counter-stories that might elucidate how systemic oppression manifests in various intersectional spaces are silenced. Suggesting that "other forms of injustice deserve higher priority" discounts Franco's right to name his oppression. Saying that Franco's story is less valuable because "other forms of injustice" deserve higher priority is, from my perspective, a form of violence. If I take Andy's response to its logical conclusion, one might also posit that my position as a white American cisgender able-bodied academic precludes my right to speak of oppression I have experienced as a homosexual and as a woman. Black professors experience oppression. Immigrant college students experience oppression. Their stories have value. It is not the place of heterosexual, white, American, cisgender men to decide whose oppression is more worthy of analysis than others', nor is it their place to decide that Franco is not "one of those people" who deserves a voice in the relative landscape of global injustice. It could be that interest convergence in STEM education (Baber, 2015) is what makes this perspective possible, as it removes the need for action towards racial equity in the field.

Second, Franco's privilege (his college degree and his international experience) was not visible to Marietta citizens. They did not defer to Franco's privilege as an educated citizen of the world. They treated him as an invader of a white American space. Quantifying the relative privilege of each party in this situation is not a fruitful analytical direction. Being a white citizen of the rural Midwest grants powerful privilege, despite the hardships that rural poverty might present.

Third, I disagree with the assumption that a valuable learning outcome for the program was to confirm to Franco that the world is unjust. I disagree with the embedded assumption that before the program, Franco experienced the American Midwest to be a just place, and that his interaction in Marietta and in Alfred's class helped reveal the injustices of reality so as to better prepare him for future battles with systemic oppression. Franco was acutely aware of injustice in rural America as a result of personal experiences with harassment and intimidation from white citizens. Injustice was not a lesson he needed to learn. Injustice is what the program should aim to resist and reform.

Fourth, I see parallels in your response with the "boys will be boys" response to bullying of LGBTQ students. "Boys will be boys" diverts focus away from systems that engender bullying. "Boys will be boys" places focus on the bullied student, claiming this experience is a "part of life" and that the best form of action is for those who are bullied to learn how to *not* be bullied. Similarly, you suggested that Franco learned a valuable lesson: "*that the rural American Midwest is not where he can be most effective.*" Is this the lesson we want to teach our students? Should the program market itself to prospective students saying that "The experiences you will have here will help you learn to take action for the environment, as long as you are white. If you are non-white, this is not where you will be most effective and it would be best if you sought education somewhere else"? This is an absurd extension, although I find it not entirely out of sync with your response.

Finally, I disagree that Alfred's critique of Franco's language and framing "*contributed to the way he prepared for successful presentations later.*" If we define "success" as *eliminating Franco's race, ethnicity, and cultural heritage and replacing it with ideas that conform to European American hegemony*, then perhaps Alfred's admonishment of Franco's performance prepared him for success. But if we define "success" as *the construction of educational and social systems where non-American perspectives are treated as an asset to a just society*, Alfred's response did not prepare Franco for success.

Fig. 3.2 My response to my advisor

unwilling to make waves in the Mennonite community and to potentially tarnish my relationship with my advisor? How did I become a model of white fragility and interest convergence? Most importantly: What could I do to rectify this?

I began this rectification by examining my role in interest convergence, white fragility, and white privilege. My advisor was my guide through my graduate program. He'd taught me how to teach undergraduate courses, write curriculum, design assessments, conduct rigorous research, and navigate complicated systems. He'd supported me through the job search process. It was not in my interest to tarnish this relationship. When my own privilege that resulted from my status as his graduate student was threatened, I passively perpetuated systemic racism and injustice because it served *me* and my goals. While at the beginning of this story I may have tried to explain this inaction through fear of conflict and my cultural Mennonite heritage, this was no longer sufficient for interpreting my unwillingness to give up my own white privilege to speak against the silencing of the experience of people of color.

I now recognize the need to face the ethical consequences of my own inaction, and that means owning the interest convergence and white privilege evident in my behavior. I continue this rectification by offering a response to my advisor's critique of my dissertation in this chapter. Although responding to these ideas publically is just a first step in a long journey to restore my commitment to dismantling injustice in science education, it is important that I do not let these conversations live and die behind closed doors in academia.

3.6 Conclusion

This exploration of my own complicit support of injustice in science education started with an examination of myself (as an activist and researcher) juxtaposed with part of my culture (a pacifist Mennonite). Although I claim a commitment to justice in science education through critical research (a source of pride), I am not proud of the moments when I have responded to injustice by *"go[ing] placidly amid the noise and haste, and remember what peace there may be in silence"* (a source of shame). My own vulnerabilities exposed through this inquiry have helped me interrogate what I previously understood to be a noble cultural commitment to pacifism and quietism (a source of pride). Instead, I now think of this pacifism and silence, in some contexts, as a means of perpetuating injustice (a source of shame). My pacifism has been a tool for the preservation of racial and social injustice in education. Any critical scholar must confront the quietist and activist selves within them to understand how their work can be used to dismantle systemic oppression of students, teachers, and scholars. In my case, this means continued reflection on how to respond to issues of systemic racism, heterosexism, and other forms of oppression through counter-narratives and stories from the field. This means not only examining my own racial identity and privilege, but also examining my place in oppressive systems. As I am certain to find myself working within these pervasive systems, identification of opportunities for resistance and action is essential. This also means continuously examining my own actions to identify interest convergence and working towards becoming an ally. There may be peace to be found in silence, but that peace is only valuable to the science education community if accompanied with confrontation of oppression. This should be a source of pride for us all.

Acknowledgements I would like to thank Dr. Django Paris for his thoughtful counsel during the preparation of this manuscript.

References

Baber, L. D. (2015). Considering the interest-convergence dilemma in STEM education. *The Review of Higher Education, 38*(2), 251–270.

Braght, T. J., & Sohm, J. F. (1987). *The bloody theater: Or, martyrs mirror of the defenseless Christians*. Scottdale: Herald Press.

Comrades-in-arms [Editorial]. (2009). *The economist*. Retrieved from http://www.economist.com/node/13877186

DeCuir, J. T., & Dixson, A. D. (2004). "So when it comes out, they aren't that surprised that it is there": Using critical race theory as a tool of analysis of race and racism in education. *Educational Researcher, 33*(5), 26–31. https://doi.org/10.3102/0013189x033005026.

Delgado, R., & Stefancic, J. (2012). *Critical race theory, an introduction*. New York: NYU Press.

Driedger, L., & Kraybill, D. B. (1994). *Mennonite peacemaking: From quietism to activism*. Scottsdale: Herald Press.

Ellis, C. (1999). Heartful autoethnography. *Qualitative Health Research, 9*(5), 669–683. https://doi.org/10.1177/104973299129122153.

Ellis, C., Adams, T. E., & Bochner, A. P. (2011). Autoethnography: An overview. *Forum: Qualitative Social Research, 12*(1), 273–290.

Hogg, C. (2009, June 6). China to show pride with gay festival. *British Broadcasting Corporation*. Retrieved from http://news.bbc.co.uk/2/hi/asia-pacific/8083672.stm

Hughes, R., & Giles, M. (2010). CRiT walking in higher education: Activating critical race theory in the academy. *Race Ethnicity and Education, 13*(1), 41–57. https://doi.org/10.1080/13613320903549685.

Jacobs, A. (2009, June 14). Gay festival in China pushes official boundaries. *The New York Times*, p. A6. Retrieved from http://www.nytimes.com/2009/06/15/world/asia/15shanghai.html?ref=world

Ladson-Billings, G. (1998). Just what is critical race theory and what's it doing in a nice field like education? *International Journal of Qualitative Studies in Education, 11*(1), 7–24. https://doi.org/10.1080/095183998236863.

Ladson-Billings, G., & Tate, W. F. (1995). Toward a critical race theory of education. *Teachers College Record, 97*(1), 47–68.

Lim, L. (2009, June 15). China's Gay Pride week mixes celebration, caution. *National Public Radio*. Retrieved from http://www.npr.org/templates/story/story.php?storyId=105405434

Liu, M. (2009, June 12). China Gay-Pride event meets obstacles. *Newsweek*. Retrieved from http://www.newsweek.com/china-gay-pride-event-meets-obstacles-80291?from=rss

Miller, H. K. (2016). Undergraduates in a sustainability semester: Models of social change for sustainability. *Journal of Environmental Education, 47*(1), 52–67.

Miller, H. K. (2018). Developing a critical consciousness of race in place-based environmental education: Franco's Story. *Environmental Education Research, 24*(6), 845–858.

Reddy, E. (2015, June 28). For Mennonites, questions of LGBT membership yields few easy answers. *National Public Radio*. Retrieved from http://www.npr.org/2015/06/28/415742055/for-mennonites-question-of-lgbt-membership-yields-few-easy-answers

Solorzano, D., Ceja, M., & Yosso, T. (2001). Critical race theory, racial microagressions, and campus racial climate: The experiences of African American college students. *The Journal of Negro Education, 69*(1), 60–73.

Weihua, C. (2009). Gay festival teaches tolerance. *China Daily*, p. 8. Retrieved from http://www.chinadaily.com.cn/cndy/2009-06/16/content_8286837.htm

Weiler, K. (1988). *Women teaching for change: Gender, class, & power*. Westport: Greenwood Publishing Group.

Yuval-Davis, N. (2012). Dialogical epistemology – an intersectional resistance to the "oppression olympics". *Gender & Society, 26*(1), 46–54. https://doi.org/10.1177/0891243211427701.

Hannah K. Miller is an Assistant Professor of Education at Northern Vermont University. Hannah's research focuses primarily on the scholarship of social change in the contexts of teacher education and environmental and sustainability education. Her current research project uses Critical Participatory Action Research (CPAR) and aims to make K-12 schools more inclusive and supportive to queer and transgender teachers.

Chapter 4
Finding a Critical Voice

Aswathy Raveendran

4.1 Journeying into Science Education

The year 2006 was a particularly difficult one. Having completed a master's in Biotechnology, I was at a crossroads where I had to decide between choosing a career of research in life sciences or changing tracks to engage in a more "interdisciplinary" approach to the sciences. Though my interest in the life sciences had not completely waned, experiences of engaging in the routine, mundane, "normal science" pursued in cancer biology research laboratories (where I had interned during my master's) had left me disillusioned. Perhaps owing to the pressure to direct research toward drug discovery, work in some of these laboratories revolved primarily around developing "savior" molecules to reverse carcinogenesis. These molecules, due to their role in the apoptopic pathway, are often studied as biological markers to understand the progression of a disease. Reductionist methods would establish the "savior" role of these molecules in controlling cell proliferation in different cell lines and cancer models. Hypothesis construction and execution of experiments were often done hastily and without paying proper heed to the complexity of the system in vivo. To illustrate this using creative nonfiction, I take the liberty to reconstruct from memory a conversation with the principal investigator (PI) of one laboratory where I had interned:

PI: I see that you have worked on cervical cancer models?

Me: Yes

PI: This is great! We have been looking for someone who can establish the role of <savior molecule X> in controlling carcinogenesis in cervical cancer models! Let us begin work right away!

A. Raveendran (✉)
BITS Pilani, Hyderabad Campus, TIFR, Mumbai, Maharashtra, India

J. Bazzul, C. Siry (eds.), *Critical Voices in Science Education Research*,
Cultural Studies of Science Education 17,
https://doi.org/10.1007/978-3-319-99990-6_4

What baffled me about this very first meeting was that the PI took no trouble to understand the molecular basis of the virally mediated nature of cervical carcinoma. He seemed interested in the carcinoma only because it served as yet another model to prove the effectiveness of his savior molecule theory. Apart from this, the relentless pressure to perform and publish, the competitiveness, and the sexism directed towards young women scientists in these laboratories, were realities that I began to perceive, and became anxious about negotiating.

It has to be said, however, that my discomfort with the science education that I received did not begin here. Perhaps a seed of doubt was sown during the final year of my master's course, when I was assigned to a project as part of an industrial biotechnology course to write a term paper on the Biopharmaceutical industry in India. The paper required me to collect information on the annual turnover and the economic value of the industry, as well as to assess the boost it would give to the GDP of the nation. While I was scouring sources for information, I chanced upon disturbing articles that reported on unethical clinical trials outsourced to third-world countries like India, and how poor and illiterate people are recruited for these trials. This involved a gross violation of the fundamental ethical code of informed consent. Determined to bring these issues to light in my paper, I wrote a section on these ethical concerns confronting the industry in addition to presenting information on its economic worth. The course instructor was surprisingly receptive and appreciative of my efforts in highlighting these aspects. This experience stayed with me, making me attentive to ways in which science and technology often work in collusion with larger, oppressive, social-political structures.

Parallelly, the personal experiences of growing up in a deeply patriarchal society coupled with access to reasonably good education and interactions with progressive family members and acquaintances, which helped make sense of these ambivalences, caused me to develop an interest in feminism. This perhaps made me more sensitive to the entrenched sexism I was faced with in science research institutes as a young woman researcher and later provoked me to interrogate gendered premises and stances ingrained in disciplinary discourses themselves.

Eventually, in 2007, I decided to move away from science and make sense of my educational experiences from a critical distance. I began to look for doctoral programmes that would give me this perspective. Eventually, I came across the interdisciplinary programme in Science Education at the Homi Bhabha Centre for Science Education.

In the narrative that unfolds next, I hope to illuminate certain epistemological challenges that I encountered as a student of science education in India, when trying to grapple with the positivism inherent in the understanding of science and the lack of sensitivity to the social and political moorings of this knowledge system in the mainstream science education discourse in India.

4.2 Critical Awakening: Introduction to the Feminist Critiques of Science

I had the good fortune of being introduced to the field of science education through a course taught by a feminist science educator, physicist and queer rights activist, Dr. Chayanika Shah, at the Tata Institute of Social Sciences in Mumbai. The course was focussed on the nature of science, and examined different perspectives on the goals and purposes of science education, while emphasising the historical and cultural embeddedness of science, and the need to teach science from this perspective. The philosophy of the course is captured in Shah (2012, p. 166):

> Scientific knowledge, like all other knowledges, (needs to be) perceived as being embedded in its context, and the process of its production (needs to be) seen as important for understanding it as the finished product. Hence, the issues in science education today are not only about how to teach in the classroom and how to understand how children learn but also about what it is that we wish to teach as science and about its methods, its processes, its practices, its exclusions and its achievements.

I particularly remember a lecture during which Chayanika recounted her experience of the women's health movement in India (Manorama and Shah 1996), of which she was a part. She dwelled on the political as well as epistemological questions that the movement had raised on the nature of contraceptive research, which is reductionist in its treatment of the menstrual cycle, and the dangerous interventions prescribed by the model to disrupt the menstrual cycle. Something about her account struck a chord with me, resonating with many questions that I was beginning to formulate. Curious to learn more, I went on to auditing a course in feminist science studies offered by Chayanika and a sociologist, Dr. Gita Chadha. It was taught as part of a postgraduate course in women's studies by the physicist-sociologist duo who, in their words, were engaged in the task of "saving and dismantling" science. Their experiences are detailed in Chayanika Shah and Gita Chadha (2011). Unlike any other class that I have been a part of, here were students who were very critical and often dismissive of science and were safe to articulate these views, unlike in conventional natural science classrooms. To me as well, this space afforded an opportunity to think about science differently. It was this course that introduced me to feminist science criticism and the inroads that it has made into various disciplines – in particular, biological sciences. A significant part of the course also dealt with introducing students to the discourses on science that are prevalent in postcolonial India. As I will elaborate later, engaging with this literature has made me conscious that positions on science and its role in society need to be articulated vis-a-vis one's own social, political and cultural context.

What followed was a period of intense reading and reflection. I found myself making connections that I never imagined I could make. These connections were made at two levels: the personal as well as epistemological. At the personal level, I made sense of the exclusion I faced as a woman in science. At the epistemological level, I realised that being a woman informed by a certain feminist political stance could also impact the science that I do. Helen Longino, in an article titled *Can there be a feminist science* (1987), elaborates this idea using examples from behavioral endocrinology research wherein she critiques the linear model that posits a causal and unidirectional relationship between hormones and behavior as being deterministic and androcentric. She then discusses alternative, interactionist models of understanding behavior that view the brain as being a self-regulating, complex and autonomous system, and wherein historical, psychological and environmental factors are taken into account when explaining behavior. Though the evidence to support both the models might be compelling, the commitment to one model or the other, she argues, is a matter of contextual values; and in her case, arises from a feminism that militates against the idea of viewing the body as a "prisoner of physiology", which the linear model prescribes. This idea, that one's standpoint, which is not just an outcome of one's gender but also of one's political commitment, could affect theory construction even within the natural sciences has been liberating, if also intimidating, for me. If it is impossible to preclude values from entering scientific theory construction, what will become of the status of truth? A challenge for me has been to find a position within the realism and antirealism debate.

At this stage, it was certainly propitious to my evolving perspectives that HBCSE has a tradition of work in Science-Technology-Society (STS) education. The STS course offered by the late Prof. Chitra Natarajan at HBCSE in 2008 covered an impressive sweep of topics that included globalization, health and agriculture. One of the first books we read as part of the course was the international bestseller, *The World is Flat* by Thomas Friedman (2005), that discusses globalization in the twentieth century. Friedman argues that information technology has transformed the world into a level playing field. When we began discussing the book, the first question that Chitra asked was: "Has it, really?" In the next series of lectures she exposed us to the dark side of globalization from the point of view of the developing world; in particular, the draconian sanctions imposed by organizations like the IMF and the World Bank in important sectors such as agriculture and health. Though explicit connections were not made on how globalisation impacts science and technology as institutions, I became aware of these aspects and found myself seeking and reading literature that explores these linkages (for example, Lyn Carter (2008)).

4.3 Locating Myself as a Science Educator

During the course work that I undertook at graduate school, I was exposed to diverse ideas and perspectives. Influential papers in science education like Derek Hodson's *Time for action: Science education for an alternative future* (2003) and Douglas Allchin's *Values in science: An educational perspective* (1999) helped me to situate the criticisms of science in science education, which I continued to revisit throughout my research.

The struggle, however, really began when I asked myself the question: *what does it mean to be a science educator in India who is at the same time critical of the dominant practices and discourses in science?* Answering this question led me to move beyond my personal experiences of disillusionment with science and my new-found love of the feminist critiques, and pay attention to the discourses on science prevalent in postcolonial Indian society. According to Chadha (2005), these discourses belong to three broad categories: liberal progressivist, leftist, and radical postcolonial.

The "liberal progressivist" view is the predominant view held by the scientific elite and the lay public alike: science is seen as a liberating epistemological force that is uncritically associated with development and progress. This view is enshrined in the idea of the *scientific temper*, popularised by the first Prime Minister of independent India, Jawaharlal Nehru. In a society struggling to overcome setbacks that were outcomes of poverty, malnutrition, and overpopulation, the exercise of a scientific attitude was seen to be the solution to all problems. Proponents of this view have lent support to large-scale science and technology based development projects and innovations such as big dams, and the Green Revolution in the agricultural sector. The "leftist" view is espoused by many who have been a part of the people's science movements in India. They too view science and technology as emancipatory and like the liberal progressivists, advocate a scientific approach to solving social problems. However, they are critical of the elitism in the agendas of modern science which makes it inaccessible to the masses. The "radical postcolonial" view criticises modern western science as being destructive in its objectification of life, seeking to bring it down from its epistemological pedestal and to validate alternative systems of knowledge, whose very rationality is questioned on the basis of golden standards laid down by modern western science establishment.

Shubha Ranganathan's (2014) article, which discusses the rationalist movement in India and its conflicts with practices like faith healing, brings out the tension between the leftist and postcolonial positions. While the liberal progressivist and the leftist positions have much in common in terms of their commitment to the "scientific world-view", the radical postcolonial view in its most polemical avatar is anti-science. Though at the epistemological level, the critiques of modern western science that emanate from radical postcolonial perspectives are important to consider, politically they might lend themselves to misappropriation by the Hindu Right in India (Nanda 1997). The challenge, therefore, has been to adopt a position that is critical, yet conscious of the benefits of a scientific worldview, in a political climate where cultural, hegemonic revivalist tendencies are attempting to recreate a "Hindu" science and technology to suit their agendas.

Unraveling these thoughts a little further, what I wish to say is that as a feminist, critical science educator, what I aim to bring into science education is a sensitivity to the historical, cultural and political embeddedness of science, a knowledge system that is marked by the standpoints of its practitioners. Furthermore, rather than transacting a taken for granted understanding of science, as a finished intellectual product, my interest lies in what is getting constructed as science in science education research. This will require a constant attempt to locate and map the knowledge represented to its "context of discovery". This does not however entail a rejection of science, or an embrace of relativism, but a view consistent with what Shah and Chadha (2011, p. 74) call a "critical, reflexive and empathetic approach" to science education at all levels. I also realise that in the current political scenario in India, I need to be wary of the rigid rationalism of the leftist and liberal progressivist views as much as the the dense intellectualism of the radical postcolonial perspectives, which may lend themselves to misappropriation.

However, bringing such a critical and reflexive standpoint into science education is difficult, as the Indian curriculum documents and textbooks largely reflect a liberal progressivist view (Raveendran and Chunawala 2015a). Consistent with this, the prevailing paradigm within science education research in India also emphasizes teaching content and processes.

4.4 Bringing the Critical Standpoint into My Work

When I set out on my doctoral journey, my inclination was to look at the gender question in science education as a PhD topic, with a focus on how construction of ability in relation to gender occurs in the science classroom. However, on reading an influential review paper by Jennie Brotman and Felicia Moore (2008) with my advisor Prof. Sugra Chunawala, which charts shifts in perspectives in the area of gender and science education, and thereafter, engaging more intensively with the feminist science studies literature, I felt that looking at gender based exclusionary practices in science classrooms, without a focus on the nature of the curriculum, and the kind of values that it transmits, would be futile. A value-free portrayal of science reflects a privileged masculine standpoint that needs to be challenged. In this section, I will describe the work I have done in the past 5 years and how my standpoint – as a third-world feminist science educator – has impacted the way I look at this area of research.

My work in science education has been directed at two levels of biology education: the higher secondary and the doctoral level. These levels represent the entry and exit points of a specialized education in science. While those who opt for higher secondary specialization in science need not necessarily take up science as a career, those pursuing a PhD would – in most cases – opt to undertake research in the area of biological sciences. At both levels, I have focused on socioscientific issues (SSI) as a means to politicize science education (Raveendran and Chunawala 2013).

Most of the existing research around SSIs is preoccupied with using these as a context to enhance science content knowledge and develop skills such as evidence evaluation, argumentation, moral reasoning, and so on (Sadler 2004). However, the excessive focus on skill development has lead to a reductionist treatment of the SSI itself and the complexities inherent in it. One particular aspect that many of the studies in this tradition have overlooked is the political component that is integral to these issues (Levinson 2013). This becomes striking in a developing country like India, pervaded by stark inequalities that are an outcome of class, caste, gender, and other structures, where stakeholders in a socioscientific controversy are rarely on a level playing field. A case in point are the many controversies around state-backed science and technology based development projects that have displaced and then failed to deliver justice to large sections of the poor and downtrodden, let alone have respectful discussions with them on whether to have these projects at all in the first place (Varughese 2012). My struggle has been to identify theoretical frameworks that acknowledge the political dimension inherent in SSIs.

After almost a year or two of floundering, I came across Ralph Levinson's (2006) framework that affords scope to bring out the political nature of these controversies. This framework not only considers evidence but parses out what is it at stake in a controversy in terms of multiple, mutually interacting levels of disagreement, arising from divergent worldviews, personal experiences and interest positions. When a socioscientific controversy is understood this way, one is able to assume multiple standpoints on it in terms of the lived experiences and vantage points of different participants, as opposed to assuming a universal structure to these controversies that are independent of the people involved and the contexts they come from.

The SSI that I used in my work with higher secondary biology students, was commercial surrogacy (assisted reproduction involving a third party, the gestational surrogate). Commercial surrogacy is a complex issue and in India, during the period of my fieldwork, the Assisted Reproductive Technology (ART) industry was largely unregulated. What is striking about this issue in a third world country like India is the involvement of women surrogates, most of whom hail from vulnerable and marginalized communities. Engaging with this kind of issue requires sensitivity to the conditions of the surrogate and a willingness to understand how the technology would impact a person who does not possess the requisite education to consider the meaning of the risk involved. In other words, I intend to highlight that stakeholders in such a controversy are from diverse socioeconomic backgrounds and understanding this requires a certain kind of *political* sensitivity. If we contrast this with comparable SSIs in the west, all stakeholders would be more or less equipped to engage in a reasonable debate, or possess the legal know-how to file a lawsuit against each other in case of a non-negotiable dispute. In India on the other hand, as Shiju Sam Varughese (2012) points out, the publics of science are diverse. These broadly include the educated elite who constitute the civil society and the marginalised who are voiceless in the political system. While the former take part in scientific controversies and are heard and engaged with; the latter are excluded by the democratic processes, lack access to the legal system, and are often displaced by big science and technology related projects.

The privileging of evidence in the "resolution" of a socioscientific controversy is yet another issue that I have encountered when engaging with the dominant paradigm of research in SSIs. This approach places scientific rationality on a high pedestal and advocates "consensus building" around socioscientific issues. However, there are alternative perspectives – Jan Alexis Nielson's (2013), for instance frames socioscientific deliberation as being about "what to do" and not "what is true". In his words "evidence should be substituted by a focus on how students articulate evidence vis-à-vis other factors in socioscientific activities" (p. 381). Scientific evidence ought to be treated as one among several other factors that are deemed important in socioscientific decision making, including worldviews, experiential knowledge, interest positions and other factors (those proposed in Levinson's framework, for example). These theoretical frameworks have thus questioned the notion of a value-neutral, detached vantage point in the deliberation of socioscientific controversies.

The study with biology doctoral students involved examining the ways in which they make sense of the issue of genetic determinism, with a focus on the value-considerations they bring in when evaluating a deterministic research claim (in behavioral science) presented in the media (Raveendran and Chunawala 2015b). In this study too, there was a specific focus on the extent to which students articulate the social and political implications of genetic determinism as a philosophy, and on whether they perceive the social "constructedness" of behavioral phenomena. Their positions were evaluated using a value-loaded critical thinking framework which combines the aspects of epistemic adequacy and logical consistency of arguments, as well as components of the critical pedagogy framework which is concerned with social emancipation and justice.

What is common to both my studies – with higher secondary and doctoral students – is the focus on students' understanding of epistemic, socio-political and ethical concerns in socioscientific issues. In both cases, concerns related to the social and political impact of science and technology has been central to the ways in which I have construed and presented the dilemmas to students as well as evaluated their responses. In the work with higher secondary students, presenting the dilemma from the perspective of surrogate mothers, as opposed to presenting it from the point of view of the technology (ART) helped in eliciting their value positions on issues of social justice and the ethics of the technology. Similarly, in the study with doctoral students, there was a specific focus, in the analysis that I have undertaken, on whether and how these students critiqued the foundational notions of genetic determinism as a philosophy, as well as the implications of such claims.

4.5 Defending Critical Work

The challenges faced when adopting critical perspectives in my work did not end with identifying appropriate theoretical frameworks. It has also been rather vexing to have to argue that self-criticality, reflexivity and sensitivity to the social, cultural and political context ought to be at the core of a good education in science.

As I pointed out, the mainstream science education discourse in India upholds a positivist view of science and emphasizes teaching the products and processes of science (Raveendran and Chunawala 2013, 2015a). Hence, my work is often viewed as located on the "boundaries" of science education research. These ideas stem from the belief that there is a "core" science education that is related to relaying content and disciplinary knowledge, and that the part which deals with values and humanistic dimensions lies at the periphery.

Such an understanding of science education is flawed in its assumptions about the nature of science. That science is value-free and questions related to values should be directed to its applications reflects, at best, an outdated view of the nature of science. This is to ignore or cast aside a whole body of work in sociology of science and feminist science studies that has unveiled the value-laden nature of scientific knowledge. Apart from this, the configurations of science are also changing and taking on dynamic new forms where new actors, locations and sites of expertise are emerging, wherein questions of values need to be taken head on. If this view of science is taken on board, then questions of values, ethics and politics will find its way back to the core of science education.

Acknowledgements I am grateful to Jesse Bazzul for making this chapter possible and for his critical engagement with my writing. I also thank Ralph Levinson for his constructive feedback.

References

Allchin, D. (1999). Values in science: An educational perspective. *Science & Education, 8*, 1–12. https://doi.org/10.1007/978-94-010-0730-6_13.

Brotman, J. S., & Moore, F. M. (2008). Girls and science: A review of four themes in the science education literature. *Journal of Research in Science Teaching, 45*(9), 971–1002. https://doi.org/10.1002/tea.20241.

Carter, L. (2008). Globalization and science education: The implications of science in the new economy. *Journal of Research in Science Teaching, 45*(5), 617–633. https://doi.org/10.1002/tea.20189.

Chadha, G. (2005). Towards an informed science criticism: The debate on science in postcolonial India. In K. Ganesh & U. Thakkar (Eds.), *Culture and the making of identity in contemporary India*. New Delhi: Sage.

Friedman, T. L. (2005). *The world is flat: A brief history of the twenty-first century*. New York: Farrar, Straus & Giroux.

Hodson, D. (2003). Time for action: Science education for an alternative future. *International Journal of Science Education, 25*(6), 645–670. https://doi.org/10.1080/09500690305021.

Levinson, R. (2006). Towards a theoretical framework for teaching controversial socio-scientific issues. *International Journal of Science Education, 28*(10), 1201–1224. https://doi.org/10.1080/09500690600560753.

Levinson, R. (2013). Practice and theory of socio-scientific issues: An authentic model? *Studies in Science Education, 49*(1), 99–116. https://doi.org/10.1080/03057267.2012.746819.

Longino, H. E. (1987). Can there be a feminist science? *Hypatia, 2*(3), 51–64.

Nanda, M. (1997). Against social de (con) struction of science: Cautionary tales from the third world. *Monthly Review, 48*(10), 1. https://doi.org/10.14452/mr-048-10-1997-03_1.

Nielsen, J. A. (2013). Delusions about evidence: On why scientific evidence should not be the main concern in socioscientific decision making. *Canadian Journal of Science, Mathematics, and Technology Education, 13*(4), 373–385. https://doi.org/10.1080/14926156.2013.845323.

Manorama, S., & Shah, C. (1996). Towards a new perspective on women's bodies: Learning and unlearning together. *Economic and Political Weekly, 31*(16–17), 35–38.

Ranganathan, S. (2014). The rationalist movement against quack healing: Critical questions. *Economic & Political Weekly, 24*(1), 13–15.

Raveendran, A., & Chunawala, S. (2013). Towards an understanding of socioscientific issues as means to achieve critical scientific literacy. In G. Nagarjuna, A. Jamakhandi, E. Sam (Eds.), *epiSTEME 5 international conference to review research on science, technology and mathematics education, conference proceedings*. Cinnamonteal, India, pp. 67–73. Retrieved from http://episteme.hbcse.tifr.res.in/index.php/episteme5/5/paper/view/130/13

Raveendran, A., & Chunawala, S. (2015a). Reproducing values: A feminist critique of reproductive health in the higher secondary biology textbook. *Indian Journal of Gender Studies, 22*(2), 194–218. https://doi.org/10.1177/0971521515578244.

Raveendran, A., & Chunawala, S. (2015b). Values in science: Making sense of biology doctoral students' critical examination of a deterministic claim in a media article. *Science Education, 99*(4), 669–695. https://doi.org/10.1002/sce.21174.

Sadler, T. D. (2004). Informal reasoning regarding socioscientific issues: A critical review ofresearch. *Journal of Research in Science Teaching, 41*(5), 513–536. https://doi.org/10.1002/tea.20009.

Shah, C. (2012). From numbers to structures: Navigating the complex terrains of science, education and feminism. *Contemporary Education Dialogue, 9*(2), 145–171. https://doi.org/10.1177/097318491200900202.

Shah, C., & Chadha, G. (2011). Teaching feminist science studies in India: An experiment. In S. Chunawala & M. Kharatmal (Eds.), *Proceedings of epiSTEME 4 – International conference to review research on science, technology and mathematics education* (pp. 144–149). India: Macmillan http://episteme4.hbcse.tifr.res.in/proceedings/strand-i-historical-philosophical-and-socio-cultural-studies-of-stm-implications-for-education/shah-and-chadha.

Varughese, S. S. (2012). Where are the missing masses? The quasi-publics and non-publics of technoscience. *Minerva, 50*(2), 239–254. https://doi.org/10.1007/s11024-012-9197-3.

Aswathy Raveendran has a PhD in Science Education from the BITS Pilani, Hyderabad Campus. She is a new entrant in the field of critical studies in science education and is interested in integrating insights from feminist science studies, political theory and postcolonial theory in science education research. She is currently a visiting faculty at the BITS Pilani, Hyderabad Campus.

Chapter 5
Stories of Hope

Ralph Levinson

5.1 Episodes of Ambivalence

As I am sitting down to write this in the garden on an unusually warm summer's day, the children next door are having a game of being 'teachers and children'. The play acting goes like this:

> 'Now what are you doing Charlie?'
> 'A drawing Miss.'
> 'Well stop it. Please get on with your work and I'll see you at the end of the lesson.'

There are any number of scenarios like this with the teacher staunching any sense of wonder and play with the prospect of punishment. I know the school the children go to because my own children went there. The teachers I came across at the school were welcoming, imaginative and gentle with children. And I would guess not that much has changed. But what remains is an image of the Gradgridian teacher: grim, deathly and oppressive.

A year or so ago I happened across an obituary in *The Guardian* newspaper of a man called Ormond Uren. I was immediately struck by this because Mr. Uren (I had known his initial was 'O' but not what it stood for) was my French teacher when I was 12, back in the 1960s. Even many years later I recall him vividly. He was a tall impressive figure in early middle age with a nonchalant bearing, as if life held no anxieties for him. He had a red open-top sports car, and all my school friends thought he was incredibly wealthy, aristocratic and that he owned a villa in the south of France. What didn't occur to us is why anyone with a villa in the south of France should be earning his keep teaching in a school in Hackney, one of the least salubrious parts of London. But the reason why I had a distinct picture of him was because he was one of the best teachers I ever had. He only taught us for a year and then left

R. Levinson (✉)
Institute of Education, University College London, London, UK
e-mail: r.levinson@ucl.ac.uk

J. Bazzul, C. Siry (eds.), *Critical Voices in Science Education Research*,
Cultural Studies of Science Education 17,
https://doi.org/10.1007/978-3-319-99990-6_5

the school but I learned far more with him than with all my later French teachers put together. He didn't allow us to speak anything other than French, and he was kind in a way that authoritative, confident and respected figures can be. I always looked forward to the sophisticated discussions that would emerge in French even at a young age. I had not seen nor heard from him until I came across his obituary.

So why was his obituary in a national newspaper? Sadly, fame isn't the privilege of all great teachers. If we had imagined that he lived some alternative glamorous life-style, his story was far more extraordinary than even we, as 12 year-olds, could have known. After an affair as a very young man with a Hungarian Countess, when he became a fluent Hungarian speaker, in addition to his proficiency in French and Spanish, he was designated to be parachuted behind German lines during WWII. But he was arrested just before the operation was due to take place, suspected of passing on information to a Soviet agent – a claim he vehemently denied until the end of his life, which was lively and rich. He then spent 4 years in prison. Because of his conviction as a 'communist' spy he was not allowed to follow an academic career for some time after his release but, as the obituary says – and here is the crushing verdict – his only option was "*unsatisfying* teaching posts" (Williams 2015) (See https://www.theguardian.com/world/2015/jul/23/ormond-uren).

I had a strong sense of betrayal, ironically in the context of spying, when I read this. People like Ormond Uren had convinced me of the good one could do by becoming a teacher. If one of the teachers I had admired found their vocation 'unsatisfying' then clearly I had understood myself and those around me rather differently than I had thought.

Perceptions of teaching, at least in the U.K., are complex, and deeply bound up with notions of social class, culture, social history, and academic snobbery. If Ormond Uren was unsatisfied ('unsatisfying' was the obituary writer's term, I don't know if it was his) then at least he had somehow achieved the pedagogic autonomy contemporary teachers in the U.K. can only dream about in the context of 'presentism' (Hargreaves 2010) and 'fabrication' (Ball 2013), a cynical compliance in pretending to achieve performative targets.

These scenarios and memories disturb me. I had chosen the vocation as a teacher and later as a teacher educator because I thought teaching could transform people's lives. But society and the media had a perception of teaching that lacked creativity or insight. Yes, I had seen one or two colleagues who were boring bullies. But many more took risks, significantly increased the aspirations of their students and were committed to a better world. Thirty years later I still recall the names and faces of the children I taught. Shaw's infamous – and ignorant – aphorism – "those who can do, those who can't teach" – reflects an ambivalence on the part of teachers, a defensiveness that teaching carries with it. But this is far from the whole story. In this semi-narrative I want to include stories of three science teachers, two of whom were part of a British Academy funded research project (Levinson and Martins 2012), we carried out with Brazilian colleagues. To some extent they reflect my own genesis and development as a science teacher, hence this is a kind of displaced narrative. Science is a subject which deepens this complexity and ambivalence because of its present-day associations with academic hierarchy, individualism (the scientist as genius), suspect notions of rigour and fairness, masculinism and instrumentalism in the egregious form of STEM education (Gough 2015).

Science practise has been unhelpfully associated with positivism – that it can be the measure of all things. However, as a young beginning science teacher I was moved, perhaps idealistically, by a science which had its origins in Marxist materialism through the works of scientists such as Lancelot Hogben and J.D. Bernal who saw in the role of scientific knowledge an emancipation for working people, destroying prejudices, obscurantism and myths. But my views changed pretty quickly after a few months in the classroom – for most pupils science was something to endure, and that was the start of a process of gradual marginalisation and resistance as a teacher.

That process put me in a difficult position. The colleagues with whom I felt closest politically had been drawn to post-modernism from which I also felt alienated: I am sceptical of its uneasy relativism, open to exploitation by authoritarians, climate change deniers and corporate expansionists, as much as by those struggling to break through the oppression of colonialist and masculinist science. On the other hand the reductionist view of science as descriptive and disengaged with social justice was equally alienating. If science was emancipatory it had both to encompass the possibilities of human agency (hence non-positivistic) but also that there were real structures and systems in the world which could be subject to processes of change and liberation through human possibilities. The Critical Realist philosophy of Roy Bhaskar (2002) which reflects this agency/structure dynamic forms the background to these narratives.

5.2 Transformers

The idea for this research project arose from a conversation I had about 5 years ago that disturbed me. A young student science teacher, Emma, whom I had supervised, came to see me at the end of her training year wondering whether she had what it takes to teach in a London state school. Unassuming and introspective, Emma had come to do a pre-service course having opted out of her medical degree. Feeling that medicine as a career was not for her, she transferred to do a biology degree where she graduated with the highest mark of her year. (This was a prestigious university with hundreds of students following this particular course). Rather than follow a postgraduate research career she then went to work for 2 years as an assistant in a school for children with learning difficulties where she decided she definitely wanted to be a teacher. It was at that point of her life that I supervised her teaching practise. Of the hundreds of pre-service teachers I had supervised – and many were very impressive – she was one of the most remarkable. She said very little in sessions, listened attentively and then towards the end, very diffidently, raised pertinent questions which none of us had thought about but seemed so obvious when she asked them. In her teaching practise she would prepare meticulously which reflected a deep respect for the students as learners. Students surpassed themselves in her lessons, it looked to me that they felt empowered by her presence. She was one of the few teachers I have witnessed where, at a number of points, some students experienced transformative, almost existential, moments: those moments which had always seemed to me as the *raison d'etre* for teaching. In one lesson with 13 year olds, in a school in a severely deprived area, she had prepared beautiful masks of a range of animals and plants, provided the students with pieces of string and asked them to use their masks to become an animal/plant and use the string to represent flows of energy in habitats. She gently encouraged them to articulate what was happening, how they would respond to competitors, what was meant by energy flows and why energy flowed in particular directions. Gradually the networks became more and more complex. At the end of the lesson a girl stood up and said: "Miss, I'd never thought about it this way. But does that mean we are *all* connected to each other". It was spoken as if this was a revelation, that what was hidden beneath the surface of the natural environment had suddenly become evident and real.

This episode also revealed something which I have observed with a few science teachers, one or two of whom have been student teachers. They raise, if implicitly, interconnected ontological, epistemological, political and moral questions. The ontological question is: Nature appears to be organised. Why should that be? The epistemological question is: How do we know this? The political question: What are the power relationships which privilege certain kinds of knowledge? The moral question is: What does that *mean* for us as conscious beings? In the case I described above it became clear to the students that there were dependent relationships in living systems. This was not obvious but it became so. The epistemological and political questions were not explicitly addressed (although the nature of interdependence represented through strings and masks alluded to these, as well as the sensitivity expressed towards student explanations and inferences) but the ontological aspect of what was revealed certainly paved the way to addressing these questions. And the moral question became core: if we are connected what does that tell us about ourselves and our responsibilities towards Nature and towards the Other? *This relationship between describing Nature and the socio-political questions which emerge through reflection was an epiphany for me.* It seemed to both address why teaching was so valuable and why science teaching in particular was so much more than a short term instrumental role in supplying the workforce.

So when Emma approached me and expressed self-doubts about her ability as a teacher I was alarmed. But while she was unassuming she was also sufficiently self-aware to know she was a good teacher. What worried her, she said, was the way schools felt they had to 'perform', all that mattered were exam results and how they competed with other schools, how schools were increasingly marketed on 'performance', and teachers' lack of autonomy. Even the way you taught was governed by imposed standards and although she had commanded enough respect from sympathetic staff to get around this as a pre-service teacher, that wasn't an environment she wanted to work in. So this led me to wonder how teachers like Emma make out in the world of metrication and 'fabrication'.

5.3 Against the Grain

Talking to friends in London and Brazil about teaching as an act of resistance we found some way forward in Marilyn Cochran Smith's work (1991), *Learning to teach against the grain* (the title expresses what we were looking at) and Roger Simon's inspiring book *Teaching against the grain: Texts for a pedagogy of possibility* (1992). These works are encapsulated in Gramsci's account of lives that are accountable for what they have achieved – not in an individualist or instrumental sense – but for their contributions to events which constitute a struggle for justice and fairness (Mayo 2008).

What we wanted to know was how do teachers who retain Gramsci's sense of the teacher's vocation, who are respected by staff and students alike, and who have stayed as teachers manage, indeed thrive in the world of STEM, performativity, outcomes and entrepreneurship. Because survival isn't good enough. So between us we identified three science teachers in London who fit the bill. All fulfilled the criteria: they had taught for at least 10 years, they were well-known to us for opposing any educational move they saw as threatening equality and fairness, and they were respected by students and colleagues alike. They were competent in widely accepted modes of teaching even when challenging the purposes of these modes, and identify and contend with their own doubts (Cochran-Smith 1991).

In doing so we interviewed the teachers through a narrative approach. What emerged were stories which ranged a continuum from despair to hope, more particularly how institutional structures had mediated their sense of agency. Here I depict two teachers at different points in the continuum of despair-hope. One who eventually succumbed to the pressures of 'performativity', the other whose institution gave him the oxygen to thrive but where he met an unexpected obstacle, but perhaps predictable in hindsight.

Sally had recently left school teaching to work in teacher development. She radiated happiness as a teacher and the more challenging the class the more at home she seemed in the classroom. She had been head of the science department and a notable campaigner for a reformed science curriculum. What was interesting for me in the light of perceptions about teaching was that she had come to teaching from industry. After a few years as a graduate trainee in industry she became bored with predominantly menial tasks and changed to teaching

> because of my enjoyment of the science, and the idea that science could be more fun, more motivating, more interesting, … that I really wanted to be the teacher.

She recalled her mother had been a primary school teacher and she had loved accompanying her on school trips. Helping children develop an interest in nature, she reflected, was what drew her back to teaching. That she could make a difference. There were three important 'i' aspects of Sally's commitment as a science teacher: identity, inquiry and integration. She retained a deep excitement about the practise of science, drawing on trips she had experienced as a child, and science had become 'part of <her> life'. Science, for Sally, was about being able to use inquiry as a means of transforming knowledge for human good. Context therefore was the hook for her: 'it's that thing about *living in the real world*, we don't live in atomic theory.' So, she recounted an incident that illustrated her commitment to inquiry-based science:

> one bright lad came in and said … A sister had died from sickle cell … I've heard people in Nigeria tell me this tree has anti-sickle cell properties. I said, "oh that's interesting, what can we do about that?" And that to me was what it turned from being kind of receivers of science to being their own little scientists … , really powerful . And then changing from people going "uhhhhh what's the solution" to "I know exactly what I need and how I'm going to make it happen".

And in order for knowledge to be used in a way that was socially progressive it had to be integrated with other disciplines.: 'I used to argue with the humanities departments that maybe they should do the same and combine the curriculum … to develop more a whole person …'

Developing a science curriculum at a national level while still a practising teacher was a great opportunity. But as the curriculum became established and then standardised under governmental reforms, the national emphasis on performativity came to affect her practise.

Sally is a collectivist and an activist. She recalled a strike when she was the Union secretary and the opposition to assessment systems.

> … tick boxes, objective testing awful. … Is there a fair way of getting kids to be graded in a normative distribution? You're always getting people at top and bottom. Unless you can say every child is allowed to achieve this. … We did have a strike and on the strike day we had a meeting, in that meeting was the best ever educational discussion we have ever had. Teachers had such brilliant ideas how they could have run it. How do we take that forward? As an erstwhile activist you hope those teachers will go with it, you know they could have worked it out. Everyone was so inspired. Everybody came, not just activists. Everyone had discussions. But the action, what we knew we were trying to defend what we knew worked … We will be collective no matter what they do to us.

Sally's view of collectivism was a long term one. But the performativity driven by top-down corporate management drove her to close doors, operate as a separate unit. She saw a new headteacher as a union-smasher and found she had few political resources, as an organiser and head of department, to counter the new management.

> As soon as you close the classroom door provided the kids come out having done their homework, actually even then perhaps you don't have it written in their books but provided they come out, they're motivated, they're not misbehaving and they're doing the tests and passing exams or whatever you can pretty much do what you want behind closed doors.

So Sally left teaching and took up a post in teacher development. That she is still able to work with her ideas in a different way is important. But towards the end of the narrative there is a sense that she found it impossible to combat an ideology that countered her philosophy of teaching. For Sally dominant political structures confounded agency.

Don tells a very different story. Growing up in a working class district in the English Midlands he was apprenticed as an electrician but his science teachers convinced him to stay on at school to complete his education and go on to study science at university. He was the first in his family to go to university. On leaving university he took a number of part-time and temporary posts ranging from teaching primary school children to teaching electronics in adult education.

There are aspects of Don's professional context which gave him possibilities not available to Sally. Don teaches at a college Centre for 16–19 year olds where the staff have the kind of autonomy rarely seen in other institutions in the state sector. There are various reasons for this, some of which I am familiar with having taught there about 20 years ago when I first met Don. The first is that since the Centre caters for students who are beyond the age of compulsory schooling it is not subject to the same tight regulations encountered by other schools. Second is the intellectual, professional and political commitment of many of the staff, an aspect which I found liberating during the 18 months I taught there. Often, over coffee between lessons, staff were discussing cosmology and the burdens of proof; Marxism and science; humanism and evolution. I found a number of my colleagues would have been considered rebels and misfits had they taught in other schools. And thirdly is a degree of consensus between management, staff and the student body about the educational role of the Centre. As Don recounts:

> There were all kinds of freedoms. There was the freedom to develop the kind of curriculum which suits the kids, and I think that was the culture of the place, that whole structure of team meetings what you were going to teach anybody and being involved in the staff meetings where clearly decisions were made to some degree. There seemed to be a structure … over which you had control and also responsibility. For the curriculum you felt responsibility as well … I don't think at the time I understood that wasn't what everybody did in every institution.

Don's interest in the history of science led him to read Galileo and the Homeric sagas. Combined with his interest in constructivism he started teaching physics using a storytelling structure based on the sagas, an approach which enabled him to illuminate the diverse theories which explained different natural phenomena.

> I … formed the notion about lesson planning and long term planning based on the structures of poetry. The notion being that if you're recounting an epic poem and you're in the oral tradition you need some kind of structure in order to remember so it helps you remember things and helps you find out when you've forgotten something

He found that initially his pedagogy met with some resistance from the students. Rather than management telling him that he had to meet performance targets, his students were concerned about examination performance. All very well being introduced to history and philosophy of physics but why were they studying content outside of the examination specifications? And, in any case, what *was* the right answer? Don understood the tensions and tried to meet their objections. His role, he explained to them, was to deepen their knowledge and understanding of physics. It was also to ensure they passed the examinations. But the pedagogic register for these two processes was not necessarily the same.

> I mean one of the things is going to the kids and saying look this isn't education, this is passing exams. … And you have that notion of a passport out is important to them but you

> have to be clear with them particularly if they are doing sciences or engineering subjects actually if they're to be successful, understand something about what it is they're doing, the pieces of paper at some stage won't get them very far … The degree to which I'm disingenuous even around that that because actually I think fundamentally what I've always been interested in is understanding the world, and allowing people to understand the world in which they live, which is not just about a job but actually make sense of the world in which they find themselves …

Making 'sense of the world' is a deeply philosophical point, difficult to make explicit in the school science curriculum. Don also understands that there are inherent contradictions in opening up topics for discussion

> … It is quite difficult with a number of kids, … where religion is important when you start to come across those deeply philosophical things about the nature of being, you can end up, and I have ended up in difficult situations. They can ask questions which you answer honestly which they don't like the answer to, God, how old the Universe is, what you think about evolution, all those things are problematic. I've had discussions about the nature of prayer and things. Through a number of those discussions, understood to be handled quite carefully if you're not to make the kids think you're not another oppressor. You are in a position of power and if you don't acknowledge your power position what appears to you to be honesty can actually rebound … I think certainly in the past few years I've been really clear with kids about talking to them about why it is I'm teaching them the way I'm teaching them and engaging them with that. I try not to burden them. I find that quite helpful.

5.4 Concluding Thoughts on a Socially Just Science

I have told part of the stories of Emma, Sally and Don because through them I recognise my own hopes and contradictions. There are commonalities between them. Emma demonstrates to the young people she teaches that scientific knowledge can lead us to understand connectivity as a natural reality as well as the social *need* for connectivity. Although Sally's pedagogy is different from Emma's, she also recognises the way in which knowledge can empower, for example, the need to lessen the pain of sickle cell anaemia. Her emphasis on integration and collectivity also reflects the importance of connectivity from a pedagogic perspective. Don helps his students see the multiplicity of ideas (often reflected through the tensions he feels as a teacher) – as opposed to a monologic epistemological hierarchy – which constitute what it is to know. How the teachers achieve their aims is mediated by the institutions in which they work, and these are moulded by broader political influences and norms. In Don's case the institution is enabler whereas Emma's and Sally's agencies are partially stifled. But what these three practitioners share is the way in which science knowledge is geared to social justice where the ideas of 'connectivity' and 'multiplicity/plurality' shine through.

References

Ball, S. J. (2013). *The education debate* (2nd ed.). Bristol: The Policy Press.
Bhaskar, R. (2002). *From science to emancipation*. London: Sage.
Cochran-Smith, M. (1991). Learning to teach against the grain. *Harvard Educational Review, 61*(3), 279–311. https://doi.org/10.17763/haer.61.3.q671413614502746.
Gough, A. (2015). STEM policy and science education: scientistic curriculum and socio-political silences. *Cultural Studies of Science Education, 10*(2), 445–458. https://doi.org/10.1007/s11422-014-9590-3.
Hargreaves, A. (2010). Presentism, individualism, and conservatism: The legacy of Dan Lortie's schoolteacher: A sociological study. *Curriculum Inquiry, 40*(1), 143–154. https://doi.org/10.1111/j.1467-873x.2009.00472.x.
Levinson, R., & Martins, I. (2012). *Political literacy as a component of science teacher training: opportunities and challenges for practise*. London: International Partnership & Mobility Scheme – The British Academy.
Mayo, P. (2008). Antonio Gramsci and his relevance for the education of adults. *Educational Philosophy and Theory, 40*(3), 418–435. https://doi.org/10.1111/j.1469-5812.2007.00357.x.
Simon, R. (1992). *Teaching against the grain: texts for a pedagogy of possibility*. New York: Bergin & Garvey.
Williams, F. (2015, July 23). Ormond Uren obituary. *The Guardian*. Retrieved from https://www.theguardian.com/world/2015/jul/23/ormond-uren

Ralph Levinson is Reader in Education at University College London Institute of Education. His research focuses on socio-political issues, chemistry education and science pedagogy.

Chapter 6
Reflections on Undergraduate Science Experiences: A Push to Science Teaching

David Segura and Olayinka Mohorn-Mintah

Emerging from a sharing stories of leaving science majors fields and choosing to teach instead of pursuing post-grad science opportunities, this chapter uses autoethnography to reflect on the common themes we experienced that helped us make sense of this transition from science to science education. Studying science at the undergraduate level is fraught with a multiple issues. Compared to other majors, students in science majors often require more years to attain degrees and endure highly competitive environments. Shirley Malcolm and Michael Feder (2016) recently highlighted the barriers students face in achieving 2 and 4-year science degrees, including the culture of science, with particular barriers that affect students of color. This culture frames learning environments within a particular sets of norms and dispositions, and how individuals should interact within science spaces. Our stories of choosing to pursue science education over science graduate and professional opportunities, speaks to the problems we saw experienced within this culture.

As students from traditionally underrepresented groups who entered science majors, our management of science identity within racialized spaces helps explain our transitions into science education as an approach to applying our science degrees in meaningful ways to us. In particular, we felt our experiences with discrimination in higher education, generated a desire to effect change within science spaces rather than through science careers. Our mentorship in science varied from nonexistent to supportive, but lead to the same eventual pursuit of science education, with our mentoring lacking a route to address the issues we experienced as science majors. Our experiences framed a desire for equity for others like us, fostering goals of improving the persistence and graduation of women and traditionally underrepresented minorities as necessary for us. As a Black female and Latino scientist, our

D. Segura (✉) · O. Mohorn-Mintah
University of Illinois, Chicago, IL, USA
e-mail: dsegur2@uic.edu

J. Bazzul, C. Siry (eds.), *Critical Voices in Science Education Research*, Cultural Studies of Science Education 17,
https://doi.org/10.1007/978-3-319-99990-6_6

negotiation of science spaces was mediated by who we are, and we hope our stories can help others persist in science.

We begin by outlining the origins of this chapter and its framing within autoethnography. This is followed by a brief outline of our life histories before studying at the undergraduate level, that we feel framed our experiences. Our experiences are then explored around dimensions of mentoring, discrimination and agency, to argue that the lack of underrepresented minorities in science may be from a disconnect between science practices and the desire to be agents of change.

6.1 Framing

Stemming from a discussion about our inspiration to enter doctoral programs in science education while employed as science teachers, this autoethnography became a group story of our experiences in our science education. Group storytelling provided an interpretive telling of significant insights or 'turning points' in our lives- or here, in our study of science. Ethnographies, and autoethnographies provide a clear method of presenting the 'voice of those on the wrong side of the power relationship,' with autoethnographies using the researchers experiences to do so (Clough 1988 in Denzin 2014). This process allowed us to recreate our motivations for studying science education, and for us, reframed the the experiences of underrepresented students of color in science fields from negative to powerful (Dyson 2007).

Like others, we also found that our autoethnographies brought to the fore issues of power, and dominating perspectives (Clough 1988 in Denzin 2014). We contend that academics are not homogenous, and while privilege exists in our positions as emerging academics, our positions are also entangled within systemic issues of power, especially as emerging academics of color. While acknowledging symbolic power gained by us as young academics, this belies the issues of power and discrimination that exist and are felt in academia, existing within realities that minority faculty experience negative bias from non-minority faculty as well as from their students (Lee and Janda 2006), or in the framing of which research matters. We hope our stories act as initial cases to provide insight for directions of research and further study (Yin 2013), even if we graduated in 2002 and 2004, and thus our experiences are now over 10 years removed.

6.2 Positionality

This section reflects on who we were, and are, when arriving at the university to study science and provides insight to interpreting the experiences presented during our undergraduate studies.

6.2.1 Olayinka

I was raised on Chicago's West Side, in a female-headed single-parent home, with tremendous support from my grandparents. My mother was young but college-educated, and could be considered middle class. My grandparents migrated to Chicago from the South during the Great Migration in the early 1950s. Growing up, I knew early on that I was "smart" and have early memories of teachers being challenged in their quest to support me academically. For example, in second grade, I spent a portion of my school day with the fourth-grade teacher for reading instruction. In third grade, was officially identified as "gifted", and was offered opportunities to skip grades twice during my K-5 educational experience.

As young as age 5, I remember accompanying my aunt, a neonatologist, to work. It was well understood that since I was "smart", I would pursue a demanding career that would command a nice salary. It was understood that an intelligent black woman, as myself, who could tolerate science would quite naturally become doctors. I was not at all passionate about science, as my elementary and secondary science instruction was uninspiring – with the exception of one high school biology course taken during my junior year. I was however passionate about being "smart" and I knew that a career in science, particularly in the health fields would provide me with economic stability and the ability to take care of myself and show the world that I was indeed "smart". Growing up as a young girl. I was bombarded with this imagery.

I attended a "good" urban high school in Chicago. It was considered very diverse and was located in a high SES neighborhood with a solid academic program which included several advanced placement courses, and a highly selective International Baccalaureate (IB) program.

6.2.2 David

I am the child of immigrants who were undocumented until the late 1990s, when they became naturalized citizens. Living in an era of ambivalence towards deportation, my parents maintained ties with Mexico, but their separation led to alienation from immigrant family in the US, and my schooling provided access to programs that helped me develop positive school identities, reinforced by my parents, that held education in high esteem, as my mother attended high school in Mexico, against her parents' wishes, and my father only received up to a third grade education. The youngest of three, I had opportunities because of the paths my brothers followed, with both entering selective enrolment schools, and benefitting from the unequal distribution of resource to these schools, when compared to other neighborhood, public schools. Schooling in selective enrollment schools that required admission exams led to educational opportunities due to advanced coursework, such as AP tracks and resources over those of neighborhood schools in Chicago, such as electives ranging from vocational to liberal arts enrichment, and also inculcation into a college-going culture with faculty that emphasized higher education as mandatory.

Additionally, in speaking with classmates in high school who did not attend selective enrollment schools, the additional coursework, lab experiences and field trips that provided enrichment for me, were not present for them, in which with selective enrolment schools acting as a method of entrenching inequity were magnified through support of selective enrolment schools. This was evident in persistence graduation rates at my high school, that were roughly 98%, when graduation rates in the entire city overall were less than 55% (Allensworth et al. 2016; Miller et al. 2002). These opportunities helped me access a guaranteed medical admission out of high school after completing any college degree, and left me with his access provided a confidence entering undergraduate science education, as well as connections to other high-achieving students in the same program, and with a duty to do well, but with little experience in dealing with inequity or discrimination in education.

6.3 Framing Our University Experiences

Below is our look into three themes identified common between us- discrimination, lack of agency and lack of mentoring to integrate agency and science. Conceptualizations of agency are based in the work of Juan Garibay (2015) and Maria Varelas et al. (2015), where agency identifies both the capacity to understand and act upon society that ranges from participating in civic society to transforming structures at micro, meso, or societal levels. For us, this meant addressing what was later perceived as racism, both in our classroom experiences, but also in the framing of the purposes of learning science. Below we explore themes we saw as important in our transition from studying science into science education.

6.4 Discrimination

Discrimination was a common theme identified, but was experienced differently by each of us. Below are examples of how we perceived discrimination in and out of our programs.

Sometimes, science classes were a haven from experiences of discrimination outside of science classes:

David: I ended up liking the large classes in science, as well as the impersonal TAs, after being called a 'pachuco gangster' in an English class by the professor, and accused of plagiarism by another in an honors elective course on classics (a requirement for participation in the University's Honors Program), who forced me to take the final 'on the spot' 3 weeks earlier than others to prove I did not plagiarize. It made me wonder how others perceived me when they saw me. I wondered even more when my Honors academic adviser suggested I not take more than one math or science class a semester- a hard task for someone majoring in biology, and who would eventually double major with chemistry. My work was never questioned in science, but I was also never pointed to as an example- actually no one was. This made me feel safe, hidden, but doing well.

Additionally, this feeling of being hidden was beneficial in avoiding conflict, and antagonism, tied to who 'should' be doing well in science, but potentially perceived as microagressions:

David: I remember my roommate being upset that he had only scored one or two points higher than me on a chemistry test since he had studied for much longer. I was proud of how well I did, but could never really understand why he was so upset. Upon reflection, I remember one of my physics classes, where students were openly hostile, much like my roommate, to someone for 'messing with the curve' by doing exceptionally well on the first exam. I remember this incident because he was an African-American male, one of the few I had seen in my science classes, and he had scored the highest score on the first exam by far, and I welled up with pride [as another minority male of color in science], much like during the experience with my roommate. It was only later that I thought the way they described him as being racialized... no one near me seemed to know his name, but they made sure to call him out as 'the black guy' messing with the curve.

Alternatively, for Olayinka, this discrimination was felt in science spaces and was overt:

Olayinka: In my junior year, an experience in my microbiology lab for microbiology majors solidified my outsider status in science. A male classmate, who appeared to be Korean-American, was assigned to work with me to complete an investigation. During the experiments and data collection, he completed the entire procedure, refusing to include any of my input. When I insisted, that he consider some of my input on the assignment, he responded by questioning my ability to properly collect and interpret the data. I was immediately infuriated and hurt, and as a result lashed out at him in anger in front of the class. I made such a scene that I was asked to leave the room, and had to be calmed down by the professor outside of the classroom. No one in the class stood up for me, even the black male student in the classroom. It was as if my emotional outburst proved that I did not belong.

Above, the recognition by/of us in science also was intertwined with racialized identities – an implied prompt that minority students should not be doing well, that fits with Olayinka's experiences. While this type of experience developed a positive science identity for us, it also created tensions between our desire to study science and how we identified as scientists. Rather than feel like we were being forced out, both our experiences showed that within and outside of science classes, discrimination was an issue to be addressed. These experiences too, made other less obvious examples apparent:

David: Our chemistry department of over 30 faculty, only had 2 female tenure-track faculty, and no underrepresented minority faculty. The graduate students (over 50) were not any more diverse, with only two underrepresented minority students, an African-American women who was viewed with intrigue because she only wanted a master's degree, and an Hispanic man who was chided for getting a scholarship (NSF fellowship) because he was Hispanic, but preferred to be referred to by his Anglicized name of 'William,' instead of 'Guillermo.'

While not overt racism, this lack of diversity in a department is telling, and represents another example of race priming related to our science identities. This lack of visible representation also created a sense of responsibility to give back in us and to persist as a visual presence for others.

Part of the complexity in reflecting on discrimination in science undergraduate programs is that persistently scholars have viewed science as a tool for accessing humanitarian and altruistic goals. Yet, for us, the culture and norms of science that espouse objectivity hide the need to apply altruism within science spaces, as opposed to as a tool. Rather than assume that we fit within Heidi Carlone and Angela Johnson's (2007) identities as scientists, these experiences suggest a more nuanced tug and pull between recognition and competence in science, while still being alienated because academic performance may vary in importance within a given context (Chang et al. 2011). Even though we were successful in science, the questioning by others as to our competence, and the lack of representation of people like us added to a set of already conflicted emotions. Studies on math and science identity have questioned whether this process involves competing identities leading to students withdrawing from spaces that cause identity conflict (Solomon et al. 2011), or differing identities becoming central and others marginalized, in order to be successful (Settles 2004), but for us alternative paths out of science were paved by a perceived lack of ways to address the problems we faced.

6.5 Lacking Social Agency

We entered science for the utilitarian goals of practicing science in medicine. As commonly seen among underrepresented minorities, positioning education as means to humanitarian goals was a common theme for us, but not emphasized in our science experiences (Grandy 1998). Whether in selecting initial majors, or changing majors, the dominant theme was a desire to use science to fulfill some social goals, whether it be in medicine, or in teaching. This alignment with larger goals was vital for our decision to stay within science majors rather than leave for other majors, but also key in our shift to science education as achieving equitable goals, rather than staying in science.

Olayinka: I initiated my science journey with the intent on being a medical doctor. I wanted to develop treatment and cures for microbial diseases. I naturally selected microbiology as a college major. I was motivated by a personal experience with chicken pox, having contracted it as a teen. I was intrigued because I wondered why I had never contracted chicken pox as a child, when I was around other children infected by the disease at different times. I also wondered why teenage onset of the disease was so severe. The movie Outbreak was released during my senior year of high school. Naturally, I dreamed of being a scientist that developed cures to diseases like Ebola. After spending two and a half years on a PhD in microbiology, I grew bored with bench science. I longed for more interaction with people, specifically young people. Working as a substitute teacher and cheerleading coach for extra money, spurred my interest in science education.

Like other underrepresented minorities studying science, we sought to study science for the benefit of others, but the actual study in science differed. Even within spaces like that of pre-medicine, we experienced competition more than learning, or any feeling of real application of science to real-world problems:

David: The character of the classes in biology and chemistry were vastly different. While in biology, students only seemed to care about their grades to be competitive, since this was part of being premedical students. But, in chemistry there was a shared identity of common struggle in trying to understand the material. There was a sort of pride in taking difficult classes- taking and speaking about PChem [Physical Chemistry] is easily relatable to other chemistry students in other universities. It actually felt that chemistry students wanted to learn the science- that chemistry was too tough to do just to be competitive for medical school- and was represented in the ways that people interacted in the classes.

Considerations of social impact were important in the definition of our research trajectories as well. Our impatience was similarly formed from a lack of action or impact while being a student, and potentially in careers. For Olayinka, to see the impact of science potentially in the future in academic science, or for David, even a lack of relevant curriculum, created a sense that science may not meet a need of affecting change in the ways that we had hoped. This was exasperated by the isolation from science faculty and TAs in the study of science. This lack of direct action as an outcome that influenced our decisions to pursue education as an application of science.

David: It was a weird feeling of loving learning about science, but having no clue when you would actually use this in the future. We knew you needed a degree to do anything related to medicine, and doing the labs were fun, but it was the weird mix of possible applications and the abstract. Like in PChem, the GC/MS was awesome for identification of compounds- from forensic applications to synthesis applications, much like the NMR [Nuclear magnetic resonance spectrometry] was, and that was awesome, but then having inorganic chem talk about MO theory without any context... I mean it was still cool, but sometimes you felt like you should be doing something that matters. When I was tutoring younger chemistry students at the university, I felt that I was actually helping people, rather than learning how I could possibly help people in the future. Learning Chemistry gave me a better chance to teach, but I liked Bio too, and could hopefully help more people by teaching both.

Much like the identities raised in Carlone and Johnson (2007) we entered science fields with a strong desire to help others. The identities that we sought to form as transformative agents, were not seen as a possibility for us within science, and instead as available through the teaching of science. This is not to say that the act of studying science was not empowering for us, as it gave us an easy transition into teaching, in high-needs areas, because of the high attrition of science teachers there.

This experience of teaching of science as an application of science also involved a demarcation of education as a pseudo-science space. Yet it also highlights the ways that science is detached in the ways we teach and the ways we do science. Science education researchers are conscious of this, and many schools, particularly in engineering fields, are restructuring how science is taught, from rote to the constructivist, but wonder if these changes capture the reasons why students that are capable and succeed in science, chose to leave. Below, we consider our experiences with mentoring, as an oft-cited predictor for persistence of underrepresented minorities in STEM.

6.6 Mentorship

Lack of perceived mentoring to traverse our needs for agency and science was the guidance missing for us. Existing within the culture of STEM, the onus felt placed on us to access advanced science education, and create opportunities for others. When taken into consideration with the experience of some faculty of color missing out on tenure or promotion due to mentoring (Griffin et al. 2013), our lack of mentoring was not surprising. Mentoring could also have helped us negotiate what we perceived as a lack of connection to agency, or discrimination, but instead:

David: For me, the only mentor that I felt really looked out for was from an education professor that also taught an honors elective seminar. He was one of the few who would listen and helped me connect my negative experiences with structural issues in education and society. Quite the opposite of the honors instructor who accused me of plagiarism, he provided the support I learned students in other majors viewed as normal. For better or worse, this cemented the idea that we just don't talk about 'these' issues that mattered to me in science, those issues that explained the reason for my struggles in school, and so spurred me on the path to be a teacher, instead of a doctor.

This lack of mentoring, fueled desire to mentor others in ways we did not receive. Specifically, this shifted the search of support from faculty/staff centered, to peer centered, in order to provide those experience that would help others like us engage with science (Strayhorn 2010).

Olayinka: Through my educational experience, I considered myself a smart and extremely capable student, even in undergrad and during my graduate studies. I knew how to "do school." In my peer circles, I was one of just a few natural science majors. While I studied regularly with a small core group of study buddies, I did not have any professional mentors or older experienced science students providing me with academic career selection support. As an undergraduate, I proudly assumed the "trailblazer" role and set out to "represent" African Americans as being successful in science. At different points in both my undergraduate and graduate school careers, I found myself mentoring and tutoring younger students, several who later earned degrees in microbiology and chemistry.

Advising, a guide to formal mentoring opportunities, sometimes caused a reminder of lowered expectations from staff and faculty, and that blunted further exploration of mentoring opportunities, or whether or not mentoring opportunities were beneficial:

David: I stopped going to my honors advisor after that first semester where I did well taking 3 courses in math and science. Hearing all the horror stories of undergraduate research- mainly of being asked to do menial tasks like wash dishes, or to clean up the lab- I was ok not taking these 'opportunities,' nor did I really feel I would benefit from them.

While mentoring can be powerful in engaging all students into science spaces, this assumes the culture of science as a space meant for all, and does not account for variation in mentoring provided. While the positive benefits of mentoring are linked to helping underrepresented minorities persist in science, it is also suggestive of how exclusive the field can be, due to limited mentoring opportunities and time investment by mentors and mentees alike. Below we discuss implications of our experiences to science education. Most notably, lack of mentoring did not afford us the type of dialogue or interactions to address the issues of discrimination or social agency, and in retrospect, might have helped us negotiate the tensions raised in the previous two sections.

Furthermore, it raises questions of identity with respect to representation. The lack of mentoring provided a clear need to connect practice of science social agency- eventually through education for social justice- as we could fill this need by mentoring others. This trend of giving back has been noted as more common among students of color once they graduate (Bowen and Bok 1999) and our experiences suggest that this comes from addressing what was lacking in our experiences. This 'giving back' in the ways that we were able, even before graduation connects directly to ideas of social agency, our desire to act on the inequities experienced by us, to change science spaces for others that come after us.

6.7 Discussion

Both of us entered doctoral programs in education focused on science trying to remediate issues perceived in science. Olayinka studies science identity development in African-American women to identify ways to better support African American women in science and David studies the way that student can be supported to persist in sciences, buttressed through peers and family networks instead, in racialized contexts. While we found that our experiences were better met in education and other fields, three larger questions were identified, around the framing of success in science, applications of science, and complexities of ethnic identity.

The first question that is still unsettled in our minds is whether or not we would be considered successful science students. While achieving degrees in science, national goals for economic advancement aim to expand the science workforce, of which we are not members. Additionally, our work in science education avoids confronting issues that exist in science professions, similar to those we experienced in our study of science. Diversity and equity should also be goals of science vocations as in science education, but this too is a complex issue, especially given evidence that many science graduates are not entering science vocations, a problem magnified for minority graduates (NSB 2014).

Additionally, while there are particular fields that are exhibiting labor shortages, there is evidence that these shortages do not necessarily afford underrepresented minorities opportunities, as underrepresented minorities continue to be represented at lower rates in science vocations when compared to their rates of graduation, although exploration into this could be due to a host of factors, ranging from issues of discrimination, to detachment from the field. Science education that focuses on social justice issues allows for explorations of these issues in ways that effect change directly, as in the case of exploring issues like local environmental pollution, or issues like racism in science. These can be places for reaching students like us while studying science, where science is a tool to address social issues. Expanding lenses of critique to and through science issues can move beyond criticism of positivism that underlie science to the use of science for social discrimination (Zuberi 2001). Indeed, limiting science critiques to those of positivistic paradigms limits the types of critiques that are leveled against science fields (Kagan 2009).

Lastly, the complexities of negotiating identity were evident in our own experiences, whether in downplaying or extolling ethnic identity, or using it as a lens for developing social agency, ethnic identity continues to be an important part to our motivations. This raises questions to how racial priming and stereotype threat are managed by successful students in science, and how perceptions of discrimination are managed in ways to reduce stress, and whether our movement out of science trajectories after achieving science degrees is indicative of the limit to the sacrifices students of color make in order to participate in science, or indicative of the ultimate success of discrimination to push us out of science spaces, raising questions as to how to create a more inclusive culture of science education, where discrimination is a continuing pressure.

References

Allensworth, E. M., Healey, K., Gwynne, J. A., & Crespin, R. (2016). High school graduation rates through two Decades of District Change: The Influence of Policies, Data Records, and Demographic shifts. Research Report. University of Chicago Consortium on School Research.

Bowen, W. G., & Bok, D. (1999). *The shape of the river*. Princeton: Princeton University.

Carlone, H. B., & Johnson, A. (2007). Understanding the science experiences of successful women of color: Science identity as an analytic lens. *Journal of Research in Science Teaching, 44*(8), 1187–1218. https://doi.org/10.1002/tea.20237.

Chang, M. J., Eagan, M. K., Lin, M. H., & Hurtado, S. (2011). Considering the impact of racial stigmas and science identity: Persistence among biomedical and behavioral science aspirants. *The Journal of Higher Education, 82*(5), 564. https://doi.org/10.1353/jhe.2011.0030.

Clough, P.T. (1988). Women writing and the life history: A reading of Toni Morrison's The Bluest Eye. (Unpublished manuscript).

Denzin, N. K. (2014). *Interpretive auto-ethnography*. Thousand Oaks: Sage.

Dyson, M. (2007). My story in a profession of stories: Auto Ethnography – an empowering methodology for educators. *Australian Journal of Teacher Education, 32*(1), 36–48. https://doi.org/10.14221/ajte.2007v32n1.3.

Garibay, J. C. (2015). STEM students' social agency and views on working for social change: Are STEM disciplines developing socially and civically responsible students? *Journal of Research in Science Teaching, 52*(5), 610–632. https://doi.org/10.1002/tea.21203.

Grandy, J. (1998). Persistence in science of high-ability minority students: Results of a longitudinal study. *Journal of Higher Education, 69*, 589–620. https://doi.org/10.2307/2649210.

Griffin, K. A., Bennett, J. C., & Harris, J. (2013). Marginalizing merit?: Gender differences in black faculty D/discourses on tenure, advancement, and professional success. *The Review of Higher Education, 36*(4), 489–512. https://doi.org/10.1353/rhe.2013.0040.

Kagan, J. (2009). *The three cultures: Natural sciences, social sciences, and the humanities in the 21st century*. New York: Cambridge University Press.

Lee, G. L., & Janda, L. (2006). Successful multicultural campus: Free from prejudice toward minority professors. *Multicultural Education, 14*(1), 27.

Malcolm, S., Feder, M., & National Academies of Sciences, Engineering, and Medicine. (2016). *Barriers and opportunities for 2-year and 4-year STEM degrees: Systemic change to support diverse student pathways*. Washington, DC: The National Academies Press. https://doi.org/10.17226/21739.

Miller, S. R., Allensworth, E. M., & Kochanek, J. R. (2002). Student performance: course taking, test scores, and outcomes: the state of chicago public high schools, 1993 to 2000. Consortium on chicago school research.

National Science Board ([NSB]US). (2014). *Revisiting the STEM workforce: A companion to science and engineering indicators 2014*.

Settles, I. H. (2004). When multiple identities interfere: The role of identity centrality. *Personality and Social Psychology Bulletin, 30*(4), 487–500. https://doi.org/10.1177/0146167203261885.

Solomon, Y., Lawson, D., & Croft, T. (2011). Dealing with 'fragile identities': Resistance and refiguring in women mathematics students. *Gender and Education, 23*(5), 565–583. https://doi.org/10.1080/09540253.2010.51227.

Strayhorn, T. L. (2010). Undergraduate research participation and STEM graduate degree aspirations among students of color. *New Directions for Institutional Research, 2010*(148), 85–93. https://doi.org/10.1002/ir.364.

Varelas, M., Tucker-Raymond, E., & Richards, K. (2015). A structure-agency perspective on young children's engagement in school science: Carlos's performance and narrative. *Journal of Research in Science Teaching, 52*(4), 516–529. https://doi.org/10.1002/tea.21211.

Yin, R. K. (2013). *Case study research: Design and methods*. New York: Sage.

Zuberi, T. (2001). *Thicker than blood: How racial statistics lie*. Minneapolis: University of Minnesota Press.

David Segura a doctoral candidate in UIC's Educational Policy Studies and a former high school science teacher. His research interests include how social capital and science identity empowers or limits students' opportunities in science.

Olayinka Mohorn-Mintah is a doctoral candidate in the Curriculum Studies program at the University of Illinois at Chicago. A former public high school science teacher, her dissertation focuses on understanding the experiences of Black female high school science teachers as they develop science identities as women of color working within traditional science education spaces.

Chapter 7
Embedding Ethics of Care into Primary Science Pedagogy: Reflections on Our Criticality

Lyn Carter, Carolina Castano Rodriguez, and Jenny Martin

7.1 Introduction

We welcome the opportunity to be part of this volume contributing to the generative discussions on criticality in science education. Querying normative science education has long been a disposition of ours collectively, and individually (see for example Carter and Martin 2017; Carter et al. 2014). Our chapter here pertains to a university-school professional development partnership promoting the *ethics of care* as simultaneously a vehicle for critiquing science education, and a pedagogical process for enacting more ethically-orientated science classes. While the domains of ethics and moral education are well considered within science education scholarship (see Lee et al. 2012), the ethics of care more specifically, as either theory, disposition or practice has to date, been underexplored. Exceptions included Lake et al. (2004) who investigated the quality of the pedagogical care relations between a teacher and her students while van Sickle and Spector (1996) reported on teachers' creative abilities to apply the scientific principle of interdependence of organisms to teach and foster an ethic of care among students. If considering the ubiquitous STEM (science, technology, engineering mathematics) movement, there is a little more work in mathematics with Nicol et al. (2010) for example, extending Hackenberg's (2005) caring relational framework to include mathematical caring and pedagogical caring (see also McCloskey 2012).

In our chapter here, we begin with an overview of the ethics of care though there is room only for a taste of some of its more salient ideas. We believe that the ethics of care accedes with, and has the potential to progress, Burke and Bazzul's (2016, p. 567) view of criticality expressed as:

L. Carter (✉) · C. Castano Rodriguez · J. Martin
Australian Catholic University, Melbourne, VIC, Australia
e-mail: lyn.carter@acu.edu.au; Carolina.CastanoRodriguez@acu.edu.au; jenny.martin@acu.edu.au

J. Bazzul, C. Siry (eds.), *Critical Voices in Science Education Research*, Cultural Studies of Science Education 17,
https://doi.org/10.1007/978-3-319-99990-6_7

> a practice (that) challenges taken-for-granted assumptions, such that educators would not be able to practice the way they did before. Critique involves the imagination of a future that cannot, at least right away, be taken up into the social order (Butler 2004, in Burke and Bazzul 2016). Social and political critique continually contests the grounds, history, and purpose of, as well as 'who' is included by, the social order.

Burke and Bazzul (2016) provide an adroit summation of the aims for our school-university professional learning partnership project where we hoped to challenge teachers' 'taken-for-granted assumptions' in science education, contesting 'its grounds, history and purpose.' Carolina's adoption and development of an ethics of care approach to her own teaching and scholarship initiated the project, with Jenny and Lyn enacting roles of critical dialogue partners and co-researchers. Having described the project, we move on to report some of Carolina reflections during the project's implementation. This approach is supported by Roseboro and Ross' (2009) view that "writing as critical educators begins with the personal (and) connects with the political" (p. 20). We conclude the chapter with a discussion of some insights that we hope will iteratively progress further the analysis and reporting of our approach to critical professional learning and pedagogy.

7.2 A Critical Approach to Science Education: The Ethics of Care

For philosopher Stan van Hooft, the concept of caring has become increasingly important in ethics in recent times. van Hooft (1995) credits social psychologist and feminist Carol Gilligan with introducing expressions of caring into moral theory in her seminal 1982 book, *In a Different Voice: Psychological Theory and Women's Development.* Together with feminist philosopher and teacher Nel Nodding's *Caring: A Feminist Approach to Ethics and Moral Education* published in 1984, Gilligan proposed an alternative approach to moral problems that has come to be known as the ethics of care or care ethics. Now widely recognised and having moved far beyond its original formulations, the ethics of care promotes the ethical value of caring sometimes as a philosophical virtue (see van Hooft 1995), and at others, a moral theory to be substituted for Kantian ethics and utilitarianism (Held 2006).

While care ethics is frequently portrayed as multiple concepts and/or practices, for Giraud and Hollin (2016), caring approaches have in common processes of relationship formation attentive to the needs of others. "Caring is fundamentally relational; a mode of engagement wherein responsibility is taken for our engagements and their effects on others" (p. 29). Care is an ethic grounded in voice, in the importance of everyone having a voice, and being heard with respect and on one's own terms (Gilligan 2011). Moreover, care ethics refuses to settle issues by delegating them to rules, guidelines and values like for example, the socio-legal construct of 'justice.' Care ethicists see the danger in the universalisation of such apparently impartial and rationally-argued constructs that can deny what it is to be entangled. In care ethics there is instead, a primacy of the particular, the local and the ordinary. As Lawrence and Maitlis (2012) suggest, the specifics of caring will depend on the concrete realities of the relationship and situation, and affection and regard, not by fixed rules normative or otherwise.

Clearly apposite in the private realm, care ethics has also been widely applied to the public and political domains. Vosman (2014) notes that political scientist Joan Tronto was the first to politicise care ethics to tackle issues, he suggests, like precariousness and the evaporation of democracy. Tronto (1993) helped move care ethics from its gender alignment with the feminine to a social practice where care is seen to be:

> a species activity that includes everything that we do to maintain, continue and repair our 'world' so that we can live in it as well as possible. That would include our bodies, ourselves, and our environment, all of which we seek to interweave in a complex, life-sustaining web (Fisher and Tronto 1990. p. 40).

Laugier (2014) contemporises the politics to see care ethics as a challenge to the neoliberal demands of autonomy, resilience and self-management, redefining the human condition instead as one of vulnerability.

Interestingly, as a relational ontology, care ethics is beginning to be explored in some of the more recent scholarship on new materialism and affect. Johns-Putra (2013) utilises the 'intra-active' unit of Barad (2007) and the 'actant' of Latour (2005) to argue a nascent new materialist vision of care. She suggests that the dynamic of carer and cared-about "all have agency and identity as they come together or, more accurately, they have agency and identity *in* their coming together" (p. 132). Hence, care is the mode by which objects become known. "Care is part of the discursive and material mesh from which objects emerge. Care—in the act of being named and purportedly exercised—emerges from and re-submerges into that mesh" (p. 134.) Giraud and Hollin (2016) argue the affective dimension of caring practices. For Koivunen (2010, p. 9), the question of affect is for many scholars:

> an opportunity for increased personal and political accountability through 'a lost language of emotion' (Middleton 1992) or a rehabilitation of 'the emotional self' (Lupton 1998), for others it reads as a possibility to move beyond the individual and personal, and to relocate critical attention from language, discourse and representations to the real

Affect then, argue Giraud and Hollin (2016), takes care beyond its relational understanding to include "quality of actors that designates their capacity to affect others and be affected in turn. Care and affect are, therefore, tightly bound" (p. 30). The affective qualities of care strengthen caring because of the affective relations between entities, in that discursive and material mesh. For Niccolini (2016), it enables an openness that moves past self-contained bodies and their histories towards emergent possibilities. As Giraud and Hollin (2016), Johns-Putra (2013) and Niccolini's (2016) work amongst others is very recent, there is clearly some interesting thinking on the horizon.

Understandably, the ethics of care has made a significant contribution to education including some recent work on affect (see for example, Zembylas 2010, 2016). Noddings (1984) is unequivocal in her belief that the "primary aim of every educational institution and of every educational effort must be the maintenance and enhancement of caring" (p.175). Beyond the specific literatures of ethical and moral education, perhaps predictably, care ethics has been explored in early childhood education (Rabin 2011), special needs (James et al. 2001), urban education (Shevalier and McKenzie 2012), teacher education (Rabin 2013; Toshalis 2012), and in other areas of educational marginalisation like queering (Benson 2008). In her much quoted 1992 book *The Challenge to Care in Schools: An Alternative Approach to Education* (revised 2005), Noddings critiques contemporary education as being too preoccupied with traditional curriculum and classroom practices that reify knowledge acquisition and cognitive performance (Fien 2003). Instead, she argues for a mutuality and reciprocity, with a learning environment grounded in caring and trusting teacher-student relationships, promoted through modelling, dialogue and practice (also Noddings 2002). In addition to strategies like cooperative learning and noncompetitive grading, she posits a curriculum organised around themes of care – caring for the self; caring for the inner circle; caring for strangers and distant others; for animals, plants, and the Earth; for the human-made world; and caring for ideas. Now known as 'centres of care,' Fien (2003) refers to these themes as widening concentric circles of care that facilitate "education for international understanding and peace; …(and) … care and compassion for non-human nature (p. 5).

Noddings' (1984, 2002, 2005, 2006) views are not without their critics though. Giraud and Hollin (2016) argue that "care's 'non-innocent' qualities (could) even implicate it in processes that might have a detrimental effect" (p. 30). For example, Roseboro and Ross (2009), Patterson and Cosart (2008), and Gerstl-Pepin (2006) are amongst the many scholars of colour in the United States who regard care ethics as colour blind. Originating as it did in white feminist world views, Patterson and Cosart (2008) argue that care ethics premises different conceptions of care than those of historically, socially and politically located Black women. Rather than the promotion of Noddings' (2005) trust, Roseboro and Ross (2009) contend that distrust and suspicion are vital manoeuvres of exposing oppression and promoting resistance. "The positionality of Black women educators makes such relationships of trust improbable when considering the systemic and structural processes in place which de-voice Black women" (p. 23). Toshalis (2012) also attests to the 'whiteness of care' and affirms a type of critical care that considers power, social location, culture and access to resources. Without this Toshalis (2012) contends, care ethics becomes yet another way of reproducing culturally dominant assumptions and socioeconomic *hierarchies* that characterise Others as deficit (also Cosart and Gordon 2006). Locating care ethics as culturally Western and white is further apparent in Nagy's (2012) acknowledgement of the strong alignment of Noddings' (2005) 'centres of care' with Judeo-Christian values. For Schultz (1998), it is in the narrative of the central and unique self positioning outwards to community that is identifiably Eurocentric and culturally insular.

Notwithstanding these very valid reservations, our adoption and adaptation of Noddings (2005) 'centres of care' within our school-university project suits our pedagogical intentions. Zembylas (2010) calls for pedagogical practices of care supportive of learning environments conducive to inclusion. He is conscious of the need to not simply learn about other cultures for example, calling instead for students to experience democracy and human rights in their daily school lives that will examine taken-for-granted realities and power relations. Zembylas (2016) goes on to argue that "all pedagogies are essentially pedagogies of emotions that are inevitably implicated in the way that knowledge operates both as a provocation of transformation and as a way of structuring emotion and affect in a particular social and political context" (p. 549). Hence, we believe that carefully implemented, the 'centres of care' enable the expansion of ways of knowing within science education, allowing us to consider how science impacts diverse cultures and social groups, as well as non-human species and natural systems. Moreover, we see the 'centres of care' not as Fien's (2003) concentric circles implying nested hierarchies but as mutually entangled and co-constituted. Such a portrayal enables the complexity of all the interactions of the natural and human made systems, and for science education to approach the same from perspectives of care and compassion (see Castano 2012a, b). Hence, an ethics of care could contribute as one response to the call for a more holistic science that values relationships, connections and caring for others (Calabrese Barton 1998).

7.3 The Ethics of Care in a School-University Partnership

Our school-university professional learning project was undertaken at St Thaddeus' primary (elementary) school during the second half of 2015. Waitoller and Artiles (2016) argue that school-university partnerships play a key role in developing teacher and school capacity. Of course, a "rather loose form of *de facto* relationship" has existed for years as schools provide universities with spaces to implement the practicum component of teacher education programs (Brady 2002, p.1). More recently, such relationships have intensified almost to the point of orthodoxy with school-based research, collaborative teaching initiatives, stakeholder professional learning, joint planning, and school-university enrichment and support likely outcomes (Brady 2002). These partnerships though, are not without their difficulties. Each setting and/or organisation is saturated with its own cultural and historical understandings, its own proclivities and its priorities and constraints. For Waitoller and Artiles (2016), there is a danger of teachers becoming both the objects and subjects of learning as they attempt to negotiate their own settings while implementing strategies developed through the university professional development workshops.

St Thaddeus is a mid-sized Catholic primary school of 350 students from Preparatory to Year 6 located in an outer north-western suburb in Melbourne designated as one of the most disadvantaged areas in the state of Victoria. Over 65% of students are located with the bottom quarter of the Index of Community Socioeducational Advantage (ICSEA), a measure that includes parental educational attainment, occupation and geographical location amongst other factors. Consequently, the school attracts additional funding for social disadvantage. A little more than 94% of students have language backgrounds other than English with large Iraqi and Vietnamese representations. The Iraqi families are predominantly Chaldean Christian refugees who have experienced significant trauma. While Chaldean Christian are part of the universal Catholic Church, they keep their own rites and traditions which can differ to those associated with Western Catholicism. In an interview for an education union newsletter about some of the key issues in working with disadvantaged students and families, the school Principal commented:

> Trauma is also a significant concern for teachers in disadvantaged areas. Trauma may be as explicit as families coming as refugees from the Middle East and Africa resettling in our communities, or it can be more subtle, families experiencing separation and/or family violence. The effect on the individual can be breakdowns in relationships/friendships, unsafe behaviour/violence, withdrawal from social interactions and a myriad of other manifestations… We try to reduce trauma through safe stable classrooms, teacher professional learning and a continued focus on building positive relationships with our families. We explore social and emotional competencies throughout our curriculum planning, we invest in staffing with an emphasis on student and staff wellbeing.

Our school-university partnership arose from interest expressed by the Principal in a prior project Carolina facilitated at St Thaddeus. Carolina placed several of her preservice teacher education students at St Thaddeus to facilitate their experience with critical pedagogical approaches to science education. Given the school's ongoing concerns around disadvantage, trauma and developmentally vulnerable students, the Principal recognised Carolina's approach to care ethics as an opportunity to further promote strategic teacher professional learning, and develop a targeted curriculum. The ethics of care was clearly consistent with St Thaddeus' socio-emotional learning framework as well as its Catholic education mission of respect, compassion and equity.

As both a research and professional learning undertaking, Carolina developed and facilitated three full day workshops at our University for eight teachers from Years 3/4 and Years 5/6. The emphasis was challenging normative views of science and science education, advocating instead for approaching embedding empathy, care, compassion, responsibility, social and ecological justice, and leadership through activism. Carolina's approach echoes Alsop's (2016) sentiments: "In imagining science education, I want to hold onto intentions, openings and arrivals, as well as residual atmospheres marked by promises of affect and science and education" (p. 552). Noddings (2005) 'centres of care' became the heuristic for the development of appropriate curriculum. This saw teachers leveraging the ideas of care for self, care for family, care for community, care for distant and different cultures, care for non-human animals, care for the natural world, care for the human-made world and care for diversity of ideas to collaboratively plan and implement lesson sequences in science for their respective classes. A full planning day held at St Thaddeus and regular team meetings with the university collaborators supported teachers during the implementation phase of the project. Throughout the workshops and meetings, the focus was on new ways of understanding, learning and teaching science oriented around connectedness rather than the mastery and manipulation of nature.

Jenny and Lyn's roles were to assist with the workshop and meetings as required as well as generate the research data. The latter comprised field notes from participant observations, recordings of meetings, collecting artefacts and completing interviews with selected participants including Carolina. Given Carolina's experience with care ethics, her insights were crucial to understanding not only the project's intent and progress but to highlight issues for further investigation. Here, we report on one of Carolina's interviews held at the conclusion of the workshops but before the implementation of the teacher developed science classes. Analysis and reporting of the data is ongoing (see for example, Castano and Martin 2015; Castano et al. 2016a, b). In the selected excerpts, Carolina recounts her own journey towards criticality and reflects on those of the teachers. While direct quotes are used where possible, some of the text has been slightly altered and punctuation added for accessibility.

7.4 Journeys Towards Criticality and the Ethics of Care: Carolina and the Teachers

Growing up in Bogotá Colombia, Carolina describes her feelings of peace and well-being from an early age when she was "connected with nature … afterall, we come from nature – it makes complete sense we should be connected with nature – it's like coming home." These affections prompted Carolina to study biology at school and university which she ultimately found disappointing:

> I faced a lot of challenges with traditional perspectives on science– it desensitises to nature – so I dreamt of a feminist perspective although at the time, I didn't know what that was – I just saw science as a vehicle to help – so I started to focus on science so I could help disadvantaged communities.

Having completed her degree in molecular biology, Carolina joined a laboratory working on adapting crops to different climates within the tropics:

> I joined them because it has so very altruistic aims and goals. But then I realised that there are so many things that are not positive in science that when we try to help one community then there are so many other organisations and things in place that use science to actually disadvantage communities, and that is when I thought I do not like what I am seeing and I won't be able to stay in this field … it is compromising my own ethics.

Carolina decided she would work to challenge instrumental and technocratic views of science and teach about ways in which science can help disadvantaged communities: "I could clearly see two sides – science as profit and science as altruist benefit." Leading educational projects with disadvantaged groups in Colombia, Ecuador and Argentina embedded Carolina's beliefs that traditional science teaching disavows the voices of the marginalised in processes of acculturation. When asked whether teaching science knowledge and skills is a way for disadvantaged groups to improve their prospects as some commentators would argue, Carolina response shows her strong commitment to criticality:

> (By doing this) We are already imposing ideas and trying to prepare people to fit into that traditional model, not preparing them to maybe challenge that model ... Ethics of care will provide a more authentic venue for them, because it gives them a voice to learn the concepts because in order to criticise something, to challenge anything, to care for their lives and their own cultural believes and background so they don't get lost within that system, they need to have an even deeper understanding of that system that will need to be challenged in the future –ethics of care provides a more authentic venue to have a deeper understanding of those concepts. If the approach is to just tell them these are the concepts they need to learn, I am really ignoring their previous understanding of their lives, science, nature, the environment … so I am already disadvantaging them even more and saying: whatever you already knew, I don't care about that. You need to learn this. It is really important to honour what they believe and learn from it.

Within the project at St Thaddeus, Carolina sees issues emerging around the teachers' journeys towards criticality and their various levels of commitment. Carolina comments:

> We are starting to see those who will try to avoid the whole thing and are not that interested … they are doing it because they have to do it. Already the most passionate one and showing lots of initiative is Jane who unfortunately is leaving us soon. … She has had some experience in animal rights and really cares about various environmental issues. For some it's going to be about their own personal journey because the thing about the ethics of care is that it challenges you to care for the views of others … and some of the teachers haven't thought about that. So they have already been challenged and confronted to think about it. They know that if they say certain things then they are actually not really caring for others' views. For some of them – it's about more than what they are going to be able to do in the class– it's quite foreign the whole (care ethics) theory and the whole aspect of seeing science in another way – so their journey will probably go as far as their own personal process – so we might not be able to see care reflected in the science classes for several years. So for someone like Brigid who says "I feel like I have to hit the wall every time I talk with the students and they don't get it" … it is clear that she is trying to change the students' views to be like hers. She understands ethics of care on an intellectual level but her normal teaching practice is so much about colonising the students. She has already expressed a lot of personal confrontation. Different to level of confrontation of Jane for example who is so open to ideas … its about the challenge of doing good things for her. Brigid is saying – I am confronted because I didn't realise that I think in this way! Its an individual journey.
>
> Then there is young Denise (a first year out teacher new to St Thaddeus) saying all the time "I don't know why we are doing this?" and maybe that is a fail (of the project) because she is someone who doesn't even question their own thoughts … querying why we are saying that and how does your own background influence her … failure might be that I am sitting here because I have to but don't see the point – someone who was not prepared to be chal-

lenged. It might reflect her lack of experience as a teacher or having it imposed on her – she hasn't had much experience in the context or working with children. The other teachers though are older and have already seen that it is so challenging to work with these children that whatever maybe able to help – they say we will try it. Denise is probably not prepared to be confronted by a theory (care ethics) when she hasn't yet been confronted by the children.

7.5 Reflections on Embedding the Ethics of Care into Primary Science Pedagogy

Our overview of the ethics of care argues its potential to work towards a criticality in science education sought by scholars like Burke and Bazzul (2016) and Calabrese Barton (1998). Although care ethics is not without its constraints, there is some interesting thinking on the horizon around new materialist approaches to care, and care and affect, that may animate the field. More specifically, we utilised Noddings (2005) 'centres of care' as pedagogical devices in our school-university partnership to facilitate the professional learning of the teachers' own criticality. Given St Thaddeus' context of disadvantage, care ethics and the 'centres of care' provided an appropriate schema consonant with Zembylas's (2010) call for pedagogical care practices supportive of inclusive learning environments. Our hope was to challenge, in Burke and Bazzul (2016) terms, the teachers "taken-for-granted assumptions," and contest the "grounds, history and purpose" of science education such that they "would not be able to practice the way they did before."

We found that the more experienced teachers were open to embracing strategies like care ethics if they could see its potential in the context to work with the children. This is not surprising given scholarship that sees teachers as bricoleurs, developing professional practices that "arise from the reflexive interactions of different types of knowledge, mediating artifacts, and methods in relation to the social contexts, cultural patterns and social actions and activities that comprise the daily events of the school" (Jenlink 2006). How engaged the teachers became though, was determined by their own personal journeys towards understanding care ethics. Only one teacher, Jane, given her own personal dispositions on animal rights and environmental advocacy could immediately envisage ways of enacting the 'centres of care' in the classroom. She conceptualised and produced a 'centres of care' diagramme of interlocking jigsaw pieces as an effective pedagogical artifact for teaching and mediating the other teachers' thinking and actions in the classroom. It would be fair to suggest that by the end of the project though, rather than having become critical pedagogues, the teachers differentially made some progress to questioning some of their 'taken-for-granted' frames of reference. This outcome is consistent with Burke and Bazzul (2016) belief that "(c)ritique involves the imagination of a future that cannot, at least right away, be taken up into the social order."

In school-university partnership projects, Waitoller and Artiles (2016) cautions about the dangers of teachers becoming both the objects and subjects of learning as they attempt to negotiate their own settings while implementing the university professional development workshop strategies. As university collaborators who didn't always evince self-awareness or find time given the project's constraints, to enact an ethics of care, this is likely the case. Gilligan (2011) argues that care is an ethic grounded in voice, in the importance of everyone having a voice, and being heard with respect and on one's own terms. Carolina notes that 'the thing about the ethics of care is that it challenges you to care for the views of others.' Our responsibility in this project was to hear Denise on her own terms and care about her views. Instead, Denise became an 'object' of our theorising as she 'failed' to understand criticality in science education and adopt care pedagogies. Moreover, we didn't encourage the teachers' criticality in questioning the care approach. While we expected them to question the "taken-for-granted assumptions" of science education, it is likely the 'centres of care' became a prescriptive curriculum planning tool. Clearly, stakeholder expectations like those of the Principal and teachers, institutional practice such as workshops, participant prejudices, and time constraints are barriers that disavow some of the altruistic intentions of critical science education work. These are amongst the continuing concerns we will explore in our ongoing writing from this project.

References

Alsop, S. (2016). Encountering science education's capacity to affect and be affected. *Cultural Studies of Science Education, 11*(3), 551–565.

Barad, K. (2007). *Meeting the universe halfway: Quantum physics and the entanglement of matter and meaning*. Durham: Duke University Press.

Benson, F. J. (2008). *Teacher educators' practice of queer-care: A necessary expansion of Noddings' model of care*. McGill University (Canada), ProQuest Dissertations Publishing.

Brady, L. (2002). School University partnerships: What do the schools want? *Australian Journal of Teacher Education, 27*(1), 1–8.

Burke, L. E. C., & Bazzul, J. (2016). Locating a space of criticality as new scholars in science education. *Cultural Studies of Science Education, 12*, 565–579. https://doi.org/10.1007/s11422-016-9735-7.

Calabrese Barton, A. (1998). *Feminist science education*. New York: Teachers College Press.

Carter, L., & Martin, J. (2017). "I had to take action straight away". Pre-service teachers' accounts of pro environmental engagement. In J. L. Bencze (Ed.), *Science & technology education promoting wellbeing for individuals, societies & environments* (pp. 523–537). Dordrecht: Springer. https://doi.org/10.1007/978-3-319-55505-8_24.

Carter, L., Castano, C., & Jones, M. (2014). Transformative learning in science education: Investigating pedagogy for action. In J. L. Bencze & S. Alsop (Eds.), *Sociopolitical activism and science & technology education* (pp. 531–546). Dordrecht: Springer.

Castano, C. (2012a). Fostering compassionate attitudes and the amelioration of aggression through a science class. *Journal of Research in Science Teaching, 49*(8), 961–986.

Castano, C. (2012b). Extending the purposes of science education: addressing violence within socio-economic disadvantaged communities. *Cultural Studies of Science Education, 7*(3), 703–718.

Castano Rodriguez, C., & Martin, J. (2015). *Final report ethics of care in science: A collaboration between Catholic Education Melbourne, Australian Catholic University and Holy Child Primary School, Dallas*. East Melbourne: Catholic Education Melbourne.

Castano, C., Martin, J., & Carter, L. (2016a). *Centres of care: A guiding principal in primary science curriculum enactment for students from low socioeconomic and refugee backgrounds*. Presented at the annual conference of the National Association for Research in Science Teaching, Baltimore. April 14–17th, 2016.

Castano, C., Martin, J., & Carter, L. (2016b). *Ethics of care in science: Enacting science curriculum that responds to the local needs of low socioeconomic communities*. Paper given at the Australian Association for Research in Education (AARE) Conference, Melbourne, November 27th –December 1st, 2016.

Cozart Conrad, S., & Gordon, J. (2006). Using womanist caring as a framework to teach social foundations. *The High School Journal, 90*(1), 9–15.

Fien, J. (2003). *Learning to care: Education and compassion*. Lecture given 15 May to the Australian School of Environmental Studies Director, Griffith University EcoCentre.

Fisher, B., & Tronto, J. (1990). Towards a feminist theory of caring. In E. Abel & M. Nelson (Eds.), *Circles of care, work and identity in women's lives* (pp. 35–62). New York: State University of New York Press.

Gerstl-Pepin, C. (2006). Utilizing an "ethic of care" in leadership preparation: Uncovering the complexity of colorblind social justice. *Journal of Educational Administration, 44*(3), 250–263.

Gilligan, C. (1982). *In a different voice: Psychological theory and women's development*. Cambridge, MA: Harvard University Press.

Gilligan, C. (2011). *Interview with Carol Gilligan*. Accessed from http://ethicsofcare.org/carol-gilligan/

Giraud, E., & Hollin, G. (2016). Care, laboratory beagles and affective utopia theory. *Theory, Culture & Society, 33*(4), 27–49.

Hackenberg, A. (2005). A model of mathematical learning and caring relations. *For the Learning of Mathematics, 25*(1), 45–51.

Held, V. (2006). *The ethics of care: Personal, political, and global*. New York: Oxford University Press.

James, P., French, P., & Cranston-Gingras, A. (2001). Ethics and special education. *Focus on Exceptional Children, 34*(1), 1–16.

Jenlink, P. M. (2006). The school leader as bricoleur: Developing scholarly practitioners for our schools. *Education Leadership Review, 7*(2), 54–69.

Johns-Putra, A. (2013). Environmental care ethics: Notes toward a new materialist critique. *symploke, 21*(1–2), 125–135.

Koivunen, A. (2010). An affective turn? Reimagining the subject of Feminist Theory. In M. Liljeström & S. Paasonen (Eds.), *Working with affect in feminist readings: Disturbing differences* (pp. 8–28). London/New York: Routledge.

Lake, V. E., Jones, I., & Dagli, U. (2004). Handle with care: Integrating caring content in mathematics and science methods classes. *Journal of Research in Childhood Education., 19*(1), 5–17.

Latour, B. (2005). *Reassembling the social: An introduction to actor-network-theory*. Oxford: Oxford University Press.

Laugier, S. (2014). *Ethics as a politics of the ordinary*. Accessed from: http://ethicsofcare.org/lecture-ethics-as-a-politics-of-the-ordinary/

Lawrence, T. B., & Maitlis, S. (2012). Care and possibility: Enacting an ethic of care through narrative practice. *Academy of Management Review, 37*(4), 641–663.

Lee, H., Chang, H., Choi, K., Kim, S.-W., & Zeidler, D. (2012). Developing character and values for global citizens: Analysis of pre-service science teachers' moral reasoning on socioscientific issues. *International Journal of Science Education, 34*(6), 925–953.

Lupton, D. (1998). *The emotional self: A sociocultural exploration*. London: Sage.

McCloskey, A. (2012). Caring in professional development projects for mathematics teachers: An example of stimulation and harmonizing. *For the Learning of Mathematics, 32*(3), 28–33.

Middleton, P. (1992). *The inward gaze: Masculinity and subjectivity in modern culture*. London: Routledge.

Nagy, A. C. (2012). Book review for Nodding, N. (2005). *The challenge to care in schools: An alternative approach to education* (2nd ed.). New York: Teachers College, Columbia University. *Journal of Research on Christian Education, 21*(1), 91–96.
Nicol, C., Novakowski, J., Ghaleband, F., & Beairsto, S. (2010). Interweaving pedagogies of care and inquiry: Tensions, dilemmas and possibilities. *Studying Teacher Education, 6*(3), 235–244.
Niccolini, A. (2016). Animate affects: censorship, reckless pedagogies, and beautiful feelings. *Gender and Education, 28*(2), 230–249. https://doi.org/10.1080/09540253.2015.1121205
Noddings, N. (1984). *Caring: A feminist approach to ethics and moral education*. Berkeley: University of California Press.
Noddings, N. (2002). *Educating moral people*. New York: Teachers College Press.
Noddings, N. (2005). *The challenge to care in schools: An alternative approach to education* (2nd ed.). New York: Teachers College, Columbia University.
Noddings, N. (2006). *Critical lessons*. New York: Cambridge University Press.
Patterson, J. A., & Conrad Cozart, S. (2008). Disrupting assumptions about race, colorblindness, and caring in educational reform and qualitative research. *The Journal of Educational Foundations, 22*(3/4), 3–12.
Rabin, C. (2011). Learning to care during storytime in the current context: Moral education from the perspective of care ethics. *Journal of Research in Childhood Education, 25*(1), 45–61.
Rabin, C. (2013). Care through authenticity: Teacher preparation for an ethic of care in an age of accountability. *The Educational Forum, 77*(3), 242–255.
Roseboro, D. L., & Ross, S. N. (2009). Care sickness: Black women educators, care theory, and a hermeneutic of suspicion. *The Journal of Educational Foundations, 23*(3/4), 19–40.
Schultz, A. (1998). Caring in schools is not enough: Community, narrative, and the limits of alterty. *Educational Theory, 48*(3), 373–393.
Shevalier, R., & McKenzie, B.-A. (2012). Culturally responsive teaching as an ethics- and care-based approach to urban education. *Urban Education, 47*(6), 1086–1105.
Toshalis, E. (2012). The rhetoric of care: Preservice teacher discourses that depoliticize, deflect, and deceive. *The Urban Review, 44*(1), 1–35.
Tronto, J. C. (1993). *Moral boundaries: A political argument for an ethic of care*. New York: Routledge.
Van Hooft, S. (1995). *Caring: an essay in the philosophy of ethics*. Niwot, CO: University Press of Colorado.
van Sickle, M., & Spector, B. (1996). Caring relationships in science classrooms: A symbolic interaction study. *Journal of Research in Science Teaching, 33*(4), 433–453.
Vosman, F. (2014). *The ethics of care: A road map*. Accessed from: http://ethicsofcare.org/lecture-the-ethics-of-care-a-road-map/
Waitoller, F. R., & Artiles, A. J. (2016). Teacher learning as curating: Becoming inclusive educators in school/university partnerships. *Teaching and Teacher Education, 59*, 360–371.
Zembylas, M. (2010). The ethic of care in globalized societies: Implications for citizenship education. *Ethics and Education, 5*(3), 233–245.
Zembylas, M. (2016). Making sense of the complex entanglement between emotion and pedagogy: Contributions of the affective turn. *Cultural Studies of Science Education, 11*(3), 539–550.

Lyn Carter Lyn's works seeks new articulations of science education valuing cultural diversity, ecological sustainability and social justice in a globalised world. Her many published journal articles and book chapters emphasises consequences the for science education of globalisation and neoliberalism, and the ways in which postcolonialism and ecological sustainability can act as counter discourses.

Carolina Castano Rodriguez is a lecturer in Science and Environmental Education and the Victoria Coordinator for the National Exceptional Teaching for Disadvantaged Schools (NETDS) Program at Australian Catholic University. She has carried out research projects in Australia, Colombia, Ecuador and Argentina, incorporating transformative learning, ethics of care and critical theories to science education within socioeconomic disadvantaged populations. Her main interest is in the link between science education, social justice and ecojustice.

Jenny Martin (previously Jenny Arnold) is a lecturer in science education at the Australian Catholic University in Melbourne. Her research is ultimately concerned with learning science for just and sustainable societies. She won the Australian Association for Research in Education 2017 Early Career Researcher Award for her paper entitled, 'Researching Preservice Teachers' Environmental Agency in Times of Post Truth and Denial'. Her research on student, teacher and preservice teacher agency using discursive psychology has been published in the *Journal of Research in Science Teaching*, *Learning Culture and Social Interaction*, the *International Journal of Science Education* and *Research in Science Education*. She has made invited contributions to forums for *Cultural Studies of Science Education* and co-guest edited a special issue on student agency in learning interactions for *Learning Culture and Social Interaction*. She is a convenor of the Sociocultural and Activity Theory Special Interest Group for the Australian Association for Research in Education and in this role facilitates mentorship and networking for higher degree students and early career researchers.

Chapter 8
Science Museums: Reflections from an Autobiographical Journey

Ana Maria Navas Iannini

In this chapter, I use the autobiographical method of currere to make meaning of what have been my professional and academic experiences with science museums. The idea of using this approach to explore my own journey with these institutions emerged a couple of years ago as an attempt to disclose relations between self, as biography, and education. Inspired by the work of educators who have used currere to deepen understanding about their own professional paths, I saw in this methodology possibilities for having critical perspectives on my current doctoral research and approaching questions that have resonated in my mind for some time now. When did my intellectual interests for science museums emerge? Why and how did I choose my current field of study? What conceptions have I confronted in my academic and professional paths regarding science museums/science education? What are the positions that I currently stand on?

My first approach to currere occurred at the beginning of my PhD program, while taking a foundational course in curriculum studies where we (graduate students) were invited to use the work of William Pinar and Madeleine Grumet for our final and reflective paper. We were invited to perform currere, and to identify and describe some roots of our own interest in curriculum studies in relation to our autobiography (Bickmore 2014). My experience in that foundational course resemble the one that Beierling et al. (2014) describe, in the sense that it was an initiation to the field of curriculum studies. In that context, I experienced currere as an attempt to cultivate the attention on the ways I participate (have participated and could participate) in the world as an educator and researcher and, also, on the ways I have shaped and have been shaped by the relations with others, with academic experiences and work.

A. M. Navas Iannini (✉)
Department of Curriculum, Teaching and Learning, Ontario Institute for Studies in Education, University of Toronto, Toronto, ON, Canada
e-mail: ana.navasiannini@mail.utoronto.ca

J. Bazzul, C. Siry (eds.), *Critical Voices in Science Education Research*, Cultural Studies of Science Education 17,
https://doi.org/10.1007/978-3-319-99990-6_8

In his work, Pinar (1981) outlines the essentiality of our individual and collective stories in shaping and situating our present projects. The awareness that emerges while our own past is reconstructed reflects the ways in which our intellectual interests function. As part of the process of understanding currere as a "self-conscious conceptualization of the temporal" (Pinar 1975, p.19), I needed to gain familiarity with the four moments of reflective inquiry that this method encompasses. The first moment is a regressive one, a turn to the biographical past. It is the stage of taking special notice of educational experiences already lived and to collect data about the past through free association (Kanu and Glor 2006). The next moment is progressive, and it implies thinking about the future (an imagined one) and bringing the attention back to where our intellectual interests are going. The next moment is the analytical, a description of the autobiographical present and an identification of ideas, fields of study and disciplines toward we are attracted to and repelled from (Pinar 1975). Finally, there is the synthetical moment, the one that requests the researcher to go back and reinterpreting the living present.

Alongside these moments of inquiry, I found the work of Grumet very useful. For Grumet (1989, p. 15) "the discovery of one's own history of wonder, engagement, and mastery may assist the teacher to make these pathways in our culture accessible to students". In this spirit, Grumet (1989) invites teachers to embrace autobiographical writing and to recover the ways in which they (us) have become "knowing subjects". Along similar lines, Yatta Kanu and Mark Glor (2006, p.104) open a space to think about currere as the "beginning point for the transformation of teachers into amateur intellectuals".

Exploring the possibilities of this self-reflection method of inquiry for educators, James Eslinger (2014) used currere and autobiographical writing to deepen understandings about his positions and points of view as a PhD student and as a teacher. Through his writings, it is possible to see how currere enabled him to examine his pedagogy as a teacher and to embrace teaching science for social justice. In another example, Jo Victoria Nicholson-Goodman (2012) used autobiographical exploration to make sense of her relations with the field of study and with the people who greatly affected her sense of the self and world.

As I mentioned before, I use currere in this paper to look at my own journey with science museums and, borrowing from Nicholson-Goodman (2012), as a way to map my intellectual interests around those institutions. In my first currere attempt, I start the regressive moment by recalling and evoking educational episodes from my childhood, such as homeschooling and then, school. By reading and reading again those passages related to my past (they were many!), I was able to employ Grumet's (1989, p.15) words, "the particular from the general in [my own] account", and to focus with intensity on some episodes I lived as an undergraduate and then a Master's student, when I first decided to embrace science education and to study science museums. From those pieces, I navigate within the other moments of reflexive inquiry (currere), looking at the positions where I take now as a doctoral student immersed in the science museum landscape, as well as where I will continue to explore in terms of research and practice.

8.1 Looking at the Past: Be Careful About What Science Exhibits Should Be!

In 2004, I moved from Colombia (my home country) to Brazil, in the spirit of exploring possibilities for a Master's program. By that time, I had finished my Bachelor in biology and I could not place myself in the apparent dichotomy of either doing field work or lab work, which seemed to be the only two professional options among colleagues. I rejected both approaches and started to wonder about the possibility of working with *biology and people*, as I internally named it. At the University of São Paulo I had the first contact with professors and students working with science education and science communication. It was exciting and refreshing for me to see how those fields were combined through graduate courses and research. Everything I experienced at the beginning of the program there was new to me. The science education research group I joined had a strong focus on museum, science centres and other informal educational settings. There, I met doctoral, masters and undergraduate students from the faculty of education and, also, members of the staff of university museums, including educators and curators. These groups of people exposed me to different angles through which science museums could be studied and, also, to theory for deepening these understandings. Science museums became a daily topic of conversation and practice and also the main focus of my research.

While I was finishing my Master's dissertation, three colleagues from the research group proposed to take a chance and work together professionally. We took the risk of opening a small business and offering some sort of educational/research services to science museums in São Paulo. Although risky, as there were no other businesses working on that specific intersection of fields, the idea was exciting as it would open possibilities for us to bring theory (some of the theories we were approaching in our own research projects) to practice. Excited about initiating a professional experience together, and being lucky to have some clients actually interested in what we were trying to do, we started with small projects, such as conducting audience research and making educational materials for visitors.

At some point, one of my business partners got a huge job opportunity to develop the contents for a museum exhibit about Brazilian space science. In our Master's program we had been discussing the role of controversy in science museums, and how the display of conflicting perspectives could help visitors identify their own positions and points of view and question the scientific enterprise. With this in mind, she shared with us her ideas about developing a display on the Brazilian satellites that had failed and the ones that had worked properly, in order to share this information with visitors and, also, to look for their responses about it. We all found her suggestion fantastic! She presented the proposal to the coordinator of the exhibit and the answer arrived pretty fast: the coordinator did not want to let visitors know that national space technology fails. She requested that the exhibit contents should clearly focus on technology that had been a success.

To some extent, what we experienced regarding the space exhibit reflects traditional ways of representing and communicating science in museum exhibits. CEOs, curators and museum educators that follow this approach tend to reinforce, through their discourses the idea of well-established scientific knowledge that only flows from the specialists to the public through a top-down communication approach. We were willing to offer a display that would eventually disrupt this grand narrative inspired by scientific success and invite visitors to reflect about the processes of science and scientific inquiry. However, these were not priorities for several science museums we worked with. Similar episodes to the space science exhibit repeated during our professional work with in the upcoming years. We were often called to create exhibits content or content for educational materials that could be "trusted", that wouldn't get us or the coordinators "into trouble", that would transmit clear and valid messages. It seemed that the idea that the public needed facts and unambiguous messages about science was much more widespread than the idea of providing spaces for exploring different perspectives in which science is embedded or for understanding science as a cultural production. Often, we struggled between balancing the job opportunity in the way the client wanted (we needed to create a portfolio for our small business!) and trying to bring our reflections (and theory) to practice.

The frustration I experienced turned some aspects of my work into a challenge to find tools for changing/challenging those "unproblematic" science scenarios. As a consultant in Brazil, I produced educational materials for museums that combined scientific knowledge with other ways of knowing such as local music, poetry and art, in an effort to reflect different perspectives and points of view on the same subject. I remember using this approach in an educational material that I developed for an exhibit about water where I introduced the topic through the lyrics of the song "Água" written by the Brazilian composer Arnaldo Antunes. Also, I conducted training courses for museum educators that contested top-down communication approaches and encouraged them to establish dialogue with visitors and honour their knowledge and perspectives, instead of offering passive instruction about specific subjects displayed in the exhibits. All those attempts revealed both possibilities and challenges to working in this field and helped me to envision what my doctoral research could focus on.

8.2 Looking at the Future: What Science Museums Could Be Open to

In this progressive stage, Pinar (1975) invites us to think about the future and the place where our intellectual interests may be going. I will make an effort to look at the issues and theories that have been meaningful for me during the doctoral program and to identify possible directions of study. While outlining the topic for my PhD research, I chose to study museums exhibits that cut across science and society and challenge top-down models of science communication (by actively inviting visitors to participate and share their positions and points of view). My decision to work on these critical exhibitions, as Erminia Pedretti (2002) named them, was strongly related to a grant research project I was asked to join when I started the program that had scientific controversies and science museums at the core. I clearly saw that project as an opportunity to challenge, revisit and question some of the facets of my professional work.

In my own research, I am interrogating the motivations of curators and educators behind critical exhibitions in Brazil and Canada and I am examining the ways in which visitors engage with them. In this process, I incorporated diverse views of scientific literacy, science communication and democratic participation in science and technology. I studied with special attention to the work of scholars who have advocated for science museums as forums (Cameron 1971/2004), places where visitors can actively dialogue (Einsiedel and Einsiedel 2004) and can experience democratic mechanisms (Bell 2008).

This is where my intellectual interests about science museums currently are, but what about possible lines of thought and action? As I near the end of my PhD experience, should I follow an academic path or do research in a science museum (outside academy)? It has been exciting for me to see that there are opportunities to conduct research (in museum contexts) related to Science Technology Engineering and Math (STEM) and underserved and underrepresented communities, exploring issues of identity, diversity and equity. I feel I am being moved to examine intersections of fields I never considered before. Unexpectedly, from this job hunting, I have been jumping into theory hunting, looking at the academic work of those who have recently considered underserved communities and equity in science museums (Dawson 2014) and engagement of youth with science in out-of-school settings (Rahm 2012). Although those topics are not directly related to my current research involving models of science communication and dimensions of visitors' engagement with critical exhibitions, I courted them while interviewing and observing teenagers from low-income neighbourhoods interacting with a Brazilian museum exhibit about teen pregnancy (I will say more about this in the next section). Communities that are part of the museum need to be considered in more active and inclusive ways. I am looking for opportunities to explore and share visitors' stories, pathways and journeys, and consider how they can help to (re)think the educational role of science museums. In this search for job opportunities (and for the theory behind), I can see refinement in my intellectual interests and, also, potential (and exciting) angles for aligning the needs of underserved communities with science museums.

8.3 Looking Closely at the present: Lessons from a Brazilian Science Exhibit

At the core of my current doctoral research is a Brazilian museum exhibit about teen pregnancy and sexual practices to which I have an emotional bond. This exhibit has strongly influenced my points of view about the directions science museums could/should be open to.

Critical on many levels, this teen pregnancy exhibit delves into issues that are difficult and sensitive, that make visitors question their own beliefs and practices (see, Navas Iannini & Pedretti 2017). One of the most interesting insights I have had regarding this exhibit is the idea that science museums can be places for experiencing empathy, for being in the place of somebody else. Through drama and role play, as a way of being in the place of others, and debates, as a way of building and sharing positions, the exhibit enables young visitors to explore their dreams for life, their views about sexual practices, and their emotions facing preconceptions about sexuality and gender (Fig.8.1). Those reflections made me go back to Derek Hodson's (2013) ideas about the role of dramatization and role play in science education as means of making (socio-scientific) issues real, to precipitate feelings about them, and to find opportunities to explore (and to care about) the other.

Through the work I have developed in this display since 2013, I have been able to examine features of science exhibits that are moving towards dialogic approaches, where reflections and perspectives of visitors are needed in order to fully experience the space. In this context, I have started to honor the role of open-ended interactions in the relations established between the exhibit, the educators and the public.

Fig. 8.1 Conversation spot that is part of the visitor experience at the Brazilian exhibit *Preventing Youth Pregnancy*. Note the informal puff seats where young visitors are invited to seat or lay down and the absence of any text, panel or object. All of them are removed from the space when the conversation is taking place

However, I have also seen that this is hard to accomplish when administrative instances impose barriers to the work of curators and educators. Topics banned from exhibitions, silenced as "inappropriate", and displays transformed in order to get a more "scientifically correct" approach are only few examples of these barriers. According to its creators this exhibit was supposed to be named "It is worth dreaming". Before the opening, the title was changed by the museum to the scientific one "Preventing youth pregnancy". I see in these barriers an attempt to keep the status quo in the relations with the public, in other words, the privileged position of the museum regarding the scientific knowledge delivered to the lay audience.

In spite of those challenges, this teen pregnancy exhibit opened a space for me to think about science museums as safe and trusthwordy environments where visitors' voices can be shared and heard. Those reflections have helped me to reimagine the role of these institutions in planting seeds for social change and considering how social inclusion can be (re)framed when other ways of knowing and feeling are at play. If science museums embrace the notion of *agora,* that Duncan Cameron (1971) envisioned some time ago, they can open themselves to include true dialogue and dissonance as part of a transformative and collective experience offered to visitors.

8.4 Between Theory and Practice "I Am Placed Together" (Pinar 1975, p.27)

Through a synthetic moment of currere, Pinar (1975) invites us to interrogate the meanings of the present and to re-enter it (Kanu and Glor 2006). In this process, I realized that the *practice* in my current doctoral experience is missing. How can I help to shorten the gap between theory and practice involving the deeper understandings I have gained about science exhibits that approach complex and critical socio-scientific themes? In what ways (besides scholarly works and events) can I use my voice and points of view to work towards the idea of science museums as places for social change? Over the past 12 years, I have been either acting as a researcher in the science museum landscape or as a professional in the museum world. I have been in the field, conceptualizing and performing the daily activities of museums and I have also been engaged in courses, academic research projects and theory that have been useful to understand and to question the role(s) and goal(s) of those institutions. I can see now, how the combination of those experiences has helped shape my positions and points of view, allowing for biographic and intellectual movement (Pinar 1975).

At this point, I feel the urge to come back to one of the commitments I made to the institutions that participated in my doctoral research: to share with them my findings and the lessons I have learned. Initially, I thought that it could be done through a report for those science museums. However, I consider now that a theory-practice resource (that could be used in workshops) could help to better navigate insights, findings and lessons. I envision this resource as structured around three main sections that reflect my current research and that also speak about new directions for science museums practices:

- *Scientific literacy and science museums*: This section includes an overview of scientific literacy perspectives that consider participation in decision making, social and environmental responsibility and engagement in socio-political action. In the suggested activity, curators and facilitators are encouraged to locate goals and expectations associated with specific exhibits within those perspectives of scientific literacy.
- *Science exhibits and socio-scientific issues:* This part of the resource centres around socio-scientific issues and examples of museum exhibits that engage with them. In the proposed activity, curators and exhibit coordinators select a display in a current exhibit that they consider approaches (or has the potential to approach) socio-scientific issues. They are encouraged, then, to analyze the features of this display and the potential for generating spaces for dissonance.
- *Science communication and the visitor experience*: This section addresses models of science communication and ways of establishing relations between the exhibit and the visitors. In the proposed activity, museum educators and facilitators read about the exhibit *Preventing Youth Pregnancy.* They are encouraged, then, to start a discussion about the ways in which educators, in that exhibit, communicate with visitors and about the barriers and possibilities that dialogic and participatory models of science communication have in their own work.

8.5 Final Thoughts

In this chapter I engaged in the process of currere to understand my journey with science museums both as a researcher and a professional in the field. Starting from a subjective and social reconstruction of episodes of my past in science, education, and, then, science museums, I went back to people and places that greatly influenced my choices and fields of study. In mapping my journey, I learned how my intellectual interests related to those institutions and my evolving biography have encouraged deeper explorations, and have allowed movement in the present, thereby enabling me to advocate for science museums as places where social change and social inclusion should occur. Recalling the words of Kanu and Glor (2006, p.112), currere was an opportunity to gain voice and to wonder how this voice could be heard inside and outside academy, and could be used "to implement transformative change".

References

Beierling, S., Buitenhuis, E., Grant, K., & Hanson, A. (2014). "Course" work: Pinar's currere as an initiation into curriculum studies. *Canadian Journal for New Scholars in Education, 5*(2), 1–9.

Bell, L. (2008). Engaging the public in technology policy. A new role for science museums? *Science Communication, 29*(3), 386–398. https://doi.org/10.1177/1075547007311971.

Bickmore, K. (2014). *CTL1000: Foundations of curriculum studies*. [Class handout]. Curriculum Studies & Teacher Development Program, CTL Department, OISE,University of Toronto.

Cameron, D. (1971/2004). The museum, a temple or the forum. In G. Anderson (Ed.),*Reinventing the museum: Historical and contemporary perspectives on the paradigm shift* (pp. 61–73). New York: Altamira Press.
Dawson, E. (2014). Equity in informal science education: Developing an access and equity framework for science museums and science centres. *Studies in Science Education, 50*(2), 209–247. https://doi.org/10.1080/03057267.2014.957558.
Einsiedel, A. A., & Einsiedel, E. F. (2004). Museums as agora: Diversifying approaches to engaging publics in research. In D. Chittenden, G. Farmelo, & B. V. Lewenstein (Eds.), *Creating connections: Museums and the public understanding of current research* (pp. 73–86). Walnut Creek: AltaMira Press.
Eslinger, J. (2014). "Don't you know only white kids like science?": Currere as critical autobiography. *Canadian Journal of New Scholars in Education, 5*(1), 1–6.
Grumet, M. R. (1989). Generations: Reconceptualist curriculum theory and teacher education. *Journal of Teacher Education, Jan-Feb, 40*(1), 13–17. https://doi.org/10.1177/002248718904000104.
Hodson, D. (2013). Don't be nervous, don't be flustered, don't be scared. Be prepared. *Canadian Journal of Science, Mathematics, and Technology Education, 13*(4), 313–331. https://doi.org/10.1080/14926156.2013.845327.
Kanu, Y., & Glor, M. (2006). Currere to the rescue? Teachers as amateur intellectuals in a knowledge society. *Journal of the Canadian Association for Curriculum Studies, 4*(2), 101–121.
Navas Iannini, A. M., & Pedretti, E. (2017). Preventing youth pregnancy: Dialogue and deliberation in a science museum exhibit. *Canadian Journal of Science Mathematics and Technology Education, 17*(4), 271–287.
Nicholson-Goodman, J. V. (2012). Mapping the doctoral journey via autobiographical consciousness: Locating self and finding voice in the academy. *Journal of Curriculum Theorizing, 28*(1), 242–259.
Pedretti, E. (2002). T. Kuhn meets T. Rex: Critical conversations and new directions in science centres and science museums. *Studies in Science Education, 37*(1), 1–41.
Pinar, W. F. (1975). The method of "Currere.". *Annual Meeting of the American Educational Research Association, 2*(1975), 19–27.
Pinar, W. (1981). "Whole, bright, deep with understanding": Issues in qualitative research and autobiographical method. *Journal of Curriculum Studies, 13*(3), 173–188. https://doi.org/10.1080/0022027810130302.
Rahm, J. (2012). Diverse urban youth's learning of science outside school in university outreach and community science programs. In F. et al. (Eds.), *Second international handbook of science education* (pp. 47–58). Dordrecht: Springer International Handbooks of Education.

Ana Maria Navas Iannini Recently completed her PhD at the Ontario Institute for Studies in Education, University of Toronto. Her research focused on museum exhibits that approach socio-scientific issues and look for public engagement. Through case studies conducted in Brazil and Canada, she examined museum staff and visitors' perspectives on exhibits about teen pregnancy and waste management.

Chapter 9
Journeys as Communicative Gestures: My Relationships with/in the Sciences

Tristan Gleason

This volume of edited work provides a space for critical stories, narratives, and voices from within the field of science education, to contest its homogeneity and conservatism. Such an opportunity feels particularly important as I begin to navigate the terrain of science education research, and seek guidance with/in my complex and multifaceted relationships with science, especially its histories and possibilities in education. The work of contestation, of challenging institutional habits and taken for granted norms, requires the construction of coalitions; it is this possibility that compels me to write, and also gives me reason to hesitate. On the one hand, I want to make public my experiences with science and science education. These experiences are inextricably linked to the criticality of my voice, and exposing them opens potential points of contact and resonance with others. Moreover, sharing the particulars of our journeys and experiences works against the ubiquitous appeal of disembodied objectivity, and towards what Donna Haraway (1988) terms *situated knowledges,* recognizing how scholarship is always already shaped by specific experiences and values. On the other hand, I worry about the seductive promise of easy transparency that often adheres to narrative research, and the assumptions of stable selves and self-evident experiences that make such transparency possible.

These concerns are at the heart of Alecia Jackson and Lisa Mazzei's (2008) critique of autoethnographic practices, including the reliance on uncomplicated narrating subjects, and an unquestioned authority of experience. Jackson and Mazzei worry that even "while autoethnographers do emphasize the constructed, partial, mediated nature of their experiences, transparent meaning that is easily understood and becomes *shared* remains the goal" (p. 303). This assumption of transparency, they argue, ignores the structures of power always already shaping our acts of tell-

T. Gleason (✉)
Moravian College, Bethlehem, PA, USA
e-mail: gleasont@moravian.edu

J. Bazzul, C. Siry (eds.), *Critical Voices in Science Education Research*,
Cultural Studies of Science Education 17,
https://doi.org/10.1007/978-3-319-99990-6_9

ing, and fails to interrogate the politics of interpretation that constantly shape and reshape our experiences. Maria Lugones (2006) extends a similar set of concerns to the particular context of coalition building, especially when such coalitions seek to coalesce at the limen, "a place where transgression of the reigning order is possible" (p. 75). Although the voices in this volume share a critical attitude and a desire to forge a shared space for reshaping the terrain of science education, Lugones warns against the presupposition that occupying such spaces results in transparency among occupants. Transparency, Lugones argues, is a promise of liberalism that upholds monologism and thus seeks to erase and silence forms of difference. Instead, she reminds us: "our journeys to the limen are different, often at odds, often at great tension… the freeing spaces where we attempt to chisel our own faces are not readily accessible to each other" (p.77). I begin by recognizing two points of tension that emerge when applying Lugones' framework to this particular project.

First, I recognize that a space marked as liminal by the authority of a discipline is distinct from the limen constructed from the concatenation of oppressive forces that Lugones describes. In other words, I don't mean to equate the margins of a discipline like science education with those that exist in societies more broadly. However, the importance of engaging in *complex communication*, rather than assuming transparency, remains. As Sandra Harding (2006) and Sylvia Wynter (2003) argue, the authority of Western Science has always emerged in collusion with such historical forces as androcentrism, colonialism, and white supremacy. Thus, although the space that this volume seeks to carve out is unique to science education, it is nonetheless shaped and necessarily related to Lugones' conception of the limen.

Second, I acknowledge that the whiteness and maleness of the body that wears me complicates the possibility of my being in any liminal space, regardless of its particular construction. However, my desire to join this space of contestation through autoethnographic practice is precisely in response to my body. In academia we have come to expect, and tacitly insist, that the voices of women and scholars of color reveal their particular locations within complex social and historical terrains. Although personal experience has become an important site of inquiry and theorization in fields like intersectionality, Patricia Hill Collins (2011) notes: "white men typically do not use their experiences in this way in doing intellectual work, such that the erasure of the social location of the intellectual becomes a significant feature of Western social theory" (p. 103).

I raise these points of tension not to resolve them, but rather to embrace the uncertainty and skepticism they invoke. My decision to narrate some of the particular experiences that provoke my desire for a place in the limen of science education research is not a means of recapturing authority. That is, I am not attempting to replace the authority of white male (dis)embodied objectivity with one of critical subjectivity, which is precisely the move that Jackson and Mazzei warn against. Rather, they suggest that autoethnographers might seek to further destabilize the truth-telling aspects of their narratives, and emphasize the unreliability and fractured nature of their narrating selves. Jackson and Mazzei suggest that one source of the unproblematic authority of these narratives resides in the phenomenological theory of experience that often accompanies autoethnographic writing: “In this paradigm, experience has already happened, is already there, and becomes cumulative and homogenizing, providing a transcendental essence on which to build consciousness and knowledge” (p. 302). Although Jackson and Mazzei “acknowledge that we remain bound to experience” (p. 304), they do not explicitly explore other theories of experience that might facilitate a less authoritative telling.

One source of a different theory of experience resides in the tradition of American pragmatism, including John Dewey’s *Experience and Nature,* which emphasizes the experimental and transactional role of experience (1973). Although the pragmatist reconstruction of experience is a topic deserving of books, not a paragraph, it shifts experience from a noun to a verb, and past experiences are not fixed objects but rather active agents in reconstructing narratives. As such, narratives that mobilize experiences do so not to uncover the truth of that past action, but to reveal the ways in which the past becomes reconstructed in the present. As Dewey reminds us, this is a process where, “The old self is put off and the new self is only forming, and the form it finally takes will depend upon the unforeseeable result of an adventure. No one discovers a new world without forsaking an old one” (1981, p. 189). Autoethnographic accounts of experience, then, become not records of past adventures but active processes of transformation and becoming.

Although Lugones does not discuss either autoethnography or pragmatism, she emphasizes the importance of giving voice to the transformative journeys that propel us into spaces of liminality as an important methodological move against the lure of transparency. She offers the work of Gloria Anzaldúa as an example, remarking that “[t]he communication of her own transformation does not make the presupposition of transparency. Instead, she offers her own transformation in a communicative gesture that enacts a complex communication” (p. 81). The telling of a story involves the unsettling of both author and audience. This is no longer about transparent meaning and shared subjectivities. As Bruno Latour (2013, p. 241, italics in original) argues: “If listeners are gripped by a piece, it is not at all because they are projecting their own pathetic subjectivity on it; it is because the work demands that they… become part of its *journey*”. So onward, to journeys as communicative gestures.

9.1 The Science of Counting

9.1.1 Roundworms

The specter of objectivity lurked above my 19 year old body as I continued my initiation into the world of laboratory science. Stare and squint as I might through the dissecting microscope, the translucent millimeter long bodies of the roundworms floating in the petri dish resisted all my efforts at quantification. In addition to their being almost invisible, whether illuminated from below, above, or both, I needed to confine my count to the worms that survived their inundation with a saline solution that I hoped would kill some, but not all, of this tiny population. Did the nearly imperceptible wiggle of the worm I currently watched mark it as alive?

As a high school student, I fell in love with biology, both for its ability to answer questions about the natural world that had long intrigued me, and for its capacity to deepen my appreciation of life's lasting mysteries. I had been particularly drawn to topics in evolution, with its emphasis on the interdependence and interconnectedness of seemingly disparate phenomena. And here I was, a sophomore in college taking a class that focused solely on the topic of evolution, and I found my love of the discipline slowly slipping away.

In particular, I was increasingly frustrated with the work we were asked to complete in the lab portion of the class. As a high school student, labs were my favorite part of learning science, where we moved back and forth between discrepant and disconcerting discoveries and the theories and ideas that helped make sense of them. In the college lab, I found the emphasis to fall squarely on the act of confirming what we already knew. My staring at roundworms was part of an assignment that charged me to design an experiment that "demonstrated" the ability of a population to change through the application of a selective pressure. Over several weeks, my experimental design would lead me to expose populations of these roundworms to an environmental condition that would kill some, but not all, of a petri dish bound population. After this exposure, I would recoup the survivors, allow them to reproduce, and then repeat, all in an effort to substantiate the theory of evolution by (un) natural selection.

To facilitate this process of counting live worms, I had marked the bottom of the petri dish with grid lines, using a permanent black marker with the thinnest nib I could find. And yet, under the magnification of the dissection scope, the thin lines ballooned into wide borders. Did that possibly alive worm on the border between the grid I counted and the one I ignored warrant counting? I looked around the neon-lit lab room, and other students were dutifully engrossed in their work. I tentatively cleared my throat several times in an effort to interrupt the person working next to me.

> "How's it going?" I asked.
> "Fine."
> "This is hard, huh?"
> "What's hard? I'm using a tally system to keep track of my count."
> "I mean, how are you deciding which worms to count in the first place? I'm not always sure which ones are alive… And what about the ones that are on the borders?"
> "You are thinking about this too much. Just count."

Counting, however, had lost its innocence. I wasn't *just* counting, I was making claims about the world, and describing a reality that no longer felt so certain. In retrospect, I wish that I could have reached forward in time and grab hold of the work of Karen Barad, of Bruno Latour, work about nature as performative (Barad 2011), about an agential rather than a descriptive realism (Barad 2007), where worlds overflow our efforts at containment and classification (Latour 1991), refusing the authoritative notion of a nature waiting patiently to be described (Latour

2004). However, such theorizing was not a part of my science education, even in the liberal arts setting. Nowhere in my training did this question of interpretation arise. Interpretation, I had been told, belonged to the humanities, or maybe those scoff-inducing "softer" sciences. In biology, we relied on data, not interpretations, and I left the lab each time feeling increasingly unmoored and adrift in a discipline I believed sought foundational truths, not observations based on conditional counts.

9.1.2 Coral Reefs

And yet, the process of becoming untethered from one discipline did prompt the exploration of others, and during the following fall of my junior year I chose to study abroad in Sri Lanka. The semester long program served approximately 20 U.S. students, and each of us lived with a local family in the city of Kandy, located in the mountainous region in the center of the island. From the 2nd day of our arrival, we studied Sinhalese, one of the languages spoken in Sri Lanka, and after an extended field trip where we studied the material culture of Sri Lanka's ancient cities, we settled into two short terms of coursework. Professors from the local university taught these courses to the American students, and during the first term I studied Sri Lankan literature and politics, and in the second, Buddhism and women and gender studies.

Although the natural sciences were not a part of the formal curriculum I chose, I remember feeling that my senses had never been so alive, fueled in part by the unique biology of Sri Lanka. Most days on my walk home from school I encountered Macaque monkeys, iridescent kingfishers, and the occasional monitor lizard, and every evening the sky filled with the flapping bodies of flying foxes. During meals I relished fruits and vegetables and spices I'd never tasted before. I remember collapsing in bed at the end of the day and giggling while the tastes and sights of the day played through my body. Even everyday animals like crows became extraordinary: I remember walking through a market one afternoon and stumbling upon a strange scene where several hopped around beneath a group of water buffalos, using their beaks to extract milk from the swollen udders that hung above their heads.

During the final month of the program students were responsible for setting up an independent study, and my host family put me in touch with a friend of theirs who worked in the department of marine biology at a university on the south coast. A research team was studying the health of local coral reef ecosystems, and they invited me to join in their work. As a lifelong lover of oceans and marine life, I was ecstatic at the opportunity to engage in science in this context, to push beyond the seemingly artificial world of college lab work and into the realm of authentic science. The authentic/artificial binary is one that today makes me shudder, and yet at that time I clung to it hungrily, hoping that this distinction would facilitate my attempts at renegotiating a relationship with a discipline I had once loved.

At first, days spent snorkeling and taking underwater pictures of reef health, performing fish counts, and zipping around on boats to collect water samples from different locations revived some of my passion. Most of the coral reefs we visited were in fairly bad shape; the rapid growth of the tourism industry, as well as increased use of fertilizers in agricultural practices, resulted in harmful algal blooms associated with coral death. Other sections of reef had become covered in sediments, preventing the photosynthetic algal symbionts from supplying the coral polyps with chemical energy. Seeing large swaths of dead coral was depressing, but it also added a sense of urgency to my idealistic notions of science as social salve. Surely, when the information from studies like the one I participated in reached the public, it would influence practice and policy. Right?

When I voiced similar ideas to the Sri Lankan scientists I worked with, they did not appear to share my optimism. I grew particularly close with a PhD student named Faisal, and one afternoon we were on our way back from a snorkeling trip in a small boat with a noisy outboard motor. The roar of the engine made conversation difficult, and we followed a rocky section of coastline back to the beach where the boats were housed. Suddenly, Faisal pointed to where a person was emerging from the ocean, pulling a large object through the small surf.

"He's collecting coral," Faisal said in response to my questioning look. As we approached the beach, I saw a large tarp spread out on the beach, and several buckets covered with fraying towels and swaths of tattered cloth. I leapt out of the boat as Faisal cut the engine and helped to slow the momentum as we ran gently aground. Several different piles of coral were assembled on the tarp—large bulky chunks in one section, and smaller piles with pieces of brain and staghorn coral. Faisal removed the outboard motor and handed it to me, along with the gas can, and he gathered up the remaining supplies while I carried my armload up to upper edge of the beach. I helped pull the boat out of the water and up out of the tidal zone. Then, Faisal walked over to the tarp and began chatting with the coral collector.

He called me over, and introduced me in Sinhala. Faisal loved to surprise other Sri Lankans with my rudimentary grasp of Sinhala, and I stumbled through my name and that I was an American student, and then quickly got lost in their rapid conversation. Soon we were huddled over one of the buckets, and Faisal pulled away the cloth to reveal a large lionfish hovering inside. He saw the surprise in my eyes. We said our goodbyes, and on our walk back to his car, Faisal tried to answer my many questions.

The man made his living from the sea. Large chunks of coral could be sold to construction companies where it was ground and used to make cement and repair roads. The ornamental pieces went to different sellers who made tourist baubles and jewelry, and the tropical fish trade was booming business. No, this was not legal. And no, there weren't very often consequences to breaking this law.

I couldn't quite believe Faisal was so calm. This, in my mind, was like an unexpected encounter between an illegal logger and an ornithologist studying the spotted owl. Did the coral collector *know* that he was causing harm to this delicate ecosystem, and one whose health already seemed so precarious? I wanted him to share in my anger. Instead, he shrugged his shoulders, quietly challenged my understanding of choice, and asked me what any of this had to do with knowledge.

When I returned back to college for my final three semesters, I dutifully finished the required courses to complete my biology major. I didn't dislike science, but I felt like the complexity of science as way of knowing and responding to the world was never addressed, except perhaps by accident on the beaches of Sri Lanka. I continued to enjoy the elegance of science, particularly in the realm of neurophysiology, but struggled to imagine what role I wanted to play in the production of scientific knowledge. I recall reading Joseph Conrad's (1963) *Nostromo* later that spring, where one of the characters discusses the difference between passion and sentiment. This distinction helped: I felt sentimental about science, but it was no longer a passion. In other words, I enjoyed reading and thinking about science. It was a compelling topic for armchair speculation, but it didn't inspire me to take action, a problem that quickly became practical.

9.1.3 The Rio Grande Valley

The practical problem is one that faces most college undergrads approaching the end of their studies—what was I going to do next? I felt fairly certain that I didn't want to work in a lab, and unsure whether the aspects of science that still sustained sentimentality would do so in the realm of a career. One afternoon I walked into the student union to grab a cup of coffee before class, and saw a large white poster hung on a bulletin board, with simple blue and red writing in the center:

DO YOU WANT TO CHANGE THE WORLD?
TEACH FOR AMERICA.

Underneath were small detachable postcards, promising more information about joining this program. I am somewhat embarrassed now by how easily I was swayed by the promise of becoming a world changer, but the tactic worked. I filled out the postcard. At least now I could tell my parents that I was thinking about what came next.

Some months later, after filling out an application and taking part in a daylong group interview process, I received a packet in the mail with information detailing my expected arrival in the Rio Grande Valley of Texas 2 weeks after I graduated, where I would be teaching middle school science. I may have managed to avoid becoming a scientist, but science, it seemed, would remain part of my identity.

The bell rang, and I hustled to gather up my photocopies and head back to class. My head was spinning with a conversation that I had refused to engage, but couldn't completely ignore. Two colleagues were discussing the latest round of statistics about district graduation rates, which explained that less than 40% of the students who graduated from the middle school where I taught ended up with a high school diploma. I was less than 2 months into my first teaching job, and while I knew that the circumstances of my students' lives here in Deep South Texas were drastically different from my experiences as a public school student in Portland, Oregon, I was also surprised to find how quickly these differences seemed to evaporate when the classroom door closed and we turned our collective attention to doing science.

My training with Teach for America had emphasized the importance of holding all students to high standards, of refusing the self-fulfilling prophecies that could follow from attending too closely to the structural conditions of students' lives. But what did it mean to extol the virtues of college, to signal to students that the work they completed in my class would help prepare them for the rigors of high school and beyond, when these realities were threatened by the hardness of numerical data?

As was often the case, I found solace in the classroom, in the revitalizing energy of students exploring themselves and the subject matter of science I sought to introduce. Yet, in moments of relative silence, I couldn't get out from underneath the weight of this statistic and I returned to the simple action of counting. Statistics and probability, I knew, detailed correlations, not causation. They were patterns of likelihood, not predestinations of fate. At the same time, the "data" said to me that less than half of the students in any of the five periods of eighth grade science I taught would graduate from high school.

So as I walked around the room, listening to students' excitement about rockets/ scientists of diverse backgrounds/the animals they investigated for research projects/ the results of a surprising experiment about yeast about aquatic plants about the color of crayfish, I couldn't help but think about numbers and counting. I continued through the tables of four students, identifying the likely high school graduates: *You, you, you, you,* and all of the sudden I was at the limit. No more "you's" fit into the container of reality crafted by this data. My hands clenched in reflex, hoping to find in them a bat I could use to smash this quantitative container that imposed such crushing limits. But no bat appeared, and anyway, I didn't know where to start swinging.

9.2 Criticality and Reconstruction

I have attempted to embrace the long tradition of writing in feminist science studies that seeks to reject what Haraway calls the god's eye view, the view that is simultaneously from everywhere and nowhere. This is not simply to accept the situatedness of my voice, of my research, of my teaching, but to begin a complex form of communication about that situation. This is not to reach for transparency; indeed, such a goal is not possible. But it is to suggest that the presupposition of a veil of opacity is equally unacceptable. We as human beings do have the opportunity to relate to one another, to engage in what Jasbir Puar (2012) terms *frictional encounters*. Such encounters seek difference rather than sameness, reject the need for commensurability, and foreground the possibility that relating to others can provoke dynamic change rather than serve as an echo chamber that further entrenches what we already assume to know. And in our political present, it is precisely the shutting down of such conversations that provoke the most worry.

I write this at a time when I am exactly halfway through my first semester as a professor of education, and I am trying to make sense of, and problematize an easy sense making, of this new identity. In particular, I wonder what it means to prepare others for the task of continuing the multiple and sometimes conflicting aims of science education. I acknowledge that the narrative I voice here offers no answers, especially to the question of why, despite the critiques I level against science and science education, I still feel compelled to engage this work. That compulsion arises from the need to continue to reconstruct as we critique, to push towards more equitable and just imaginings and doings of science education while continuing to confront the ways that schooling can also be a subtractive process (Valenzuela 1999), and a site of suffering (Dumas 2014).

If we seek to escape the trap of ideological critique, what choice do we have but to lay bare those experiences that play a role in enabling us to see and tell in ways that run counter to tradition, the status quo, accepted and expected wisdom? As if theory and introspective thought alone are sufficient sources of problematizing that which we think we know?

My narrative is an attempt to share the path that has led me to a place of productive criticality, and to show that critique is not an open rejection of something perceived from afar, but rather something that develops out of closeness and appreciation. To show that education is not simply about preparation and apprenticeship, but also about coming to know something so deeply that we cannot ignore its foibles and inconsistencies. This is true both of the narrating "I" and the subject matter I teach. It also demonstrates that critique stems from love, from wanting to engage the complexities of phenomena that are both problematic and promising, or perhaps problematic because they are promising. That appreciation for a subject matter does not require dogmatic acceptance. And thus, teaching is not about transmitting dogma, or dogmatic acceptance, especially not our own, but about providing opportunities for the development of complex relationships.

I think the god's eye view is always a particularly strong temptation for bodies like mine, those whose whiteness and maleness serve to suggest a sense of universality even when we don't claim it. That is, we continue to expect scholars of color and women to account for their particular perspectives, to qualify their claims, while holding onto the seemingly normal and comprehensive experiences of the white male body. Autoethnography can become a means for resisting this temptation, and for opening up the narrating self and concomitant experiences as sites of inquiry. Rather than supply neat and tidy answers, I hope that this work prompts questions about the values and purposes of continuing to work the ruins of science education. Science is fundamentally about the messy relationships between experiences of the world and claims to knowledge and power. If we are to move towards a more complex understanding of this relationship, then surely science education has an important role to play.

References

Barad, K. (2007). *Meeting the universe halfway: Quantum physics and the entanglement of matter and meaning*. Durham: Duke University Press. https://doi.org/10.1215/9780822388128.
Barad, K. (2011). Nature's queer performativity. *Qui Parle: Critical Humanities and Social Sciences, 19*, 121–158. https://doi.org/10.5250/quiparle.19.2.0121.
Collins, P. H. (2011). Piecing together a genealogical puzzle: Intersectionality and American pragmatism. *European Journal of Pragmatism and American Philosophy, 3*, 88–112.
Conrad, J. (1963). *Nostromo: A tale of the seaboard*. Harmondsworth: Penguin Books.
Dewey, J. (1973). The need for a recovery of philosophy. In J. McDermott (Ed.), *The philosophy of John Dewey* (pp. 58–97). Chicago: University of Chicago Press.
Dewey, J. (1981). *Experience and nature*. LW (Vol. 1). Carbondale: Southern Illinois University Press.
Dumas, M. J. (2014). 'Losing an arm': Schooling as a site of black suffering. *Race Ethnicity and Education, 17*, 1–29. https://doi.org/10.1080/13613324.2013.850412.
Haraway, D. J. (1988). Situated knowledges: The science question in feminism and the privilege of partial perspective. *Feminist Studies, 14*, 575–599. https://doi.org/10.2307/3178066.
Harding, S. (2006). *Science and social inequality: Feminist and postcolonial issues*. Urbana: University of Illinois Press.
Jackson, A. Y., & Mazzei, L. M. (2008). Experience and "I" in autoethnography: A deconstruction. *International Review in Qualitative Research, 1*(3), 299–318.
Latour, B. (1991). *We have never been modern*. Cambridge, MA: Harvard University Press.
Latour, B. (2004). *Politics of nature: How to bring the sciences into democracy*. Cambridge, MA: Harvard University Press.
Latour, B. (2013). *An inquiry into modes of existence: An anthropology of the moderns*. Cambridge, MA: Harvard University Press.
Lugones, M. (2006). On complex communication. *Hypatia, 21*, 75–85. https://doi.org/10.1111/j.1527-2001.2006.tb01114.x.
Puar, J. (2012). "I would rather be a cyborg than a goddess": Becoming-intersectional in assemblage theory. *Philosophia, 2*, 49–66.
Valenzuela, A. (1999). *Subtractive schooling: U.S.-Mexican youth and the politics of caring*. Albany: State University of New York Press.
Wynter, S. (2003). Unsettling the coloniality of being/power/truth/freedom: Towards the human, after man, its overrepresentation—An argument. *CR: The New Centennial Review, 3*, 257–337. https://doi.org/10.1353/ncr.2004.0015.

Tristan Gleason I grew up in Portland, Oregon, and attended outstanding public schools that shaped the educator I am today. I earned a BA in biology from Bowdoin College in Brunswick, ME, and after graduating spent 8 years teaching middle and high school science in the Rio Grande Valley of Texas and in Portland, Oregon. My experiences as a science educator led me to pursue a PhD in Critical and Sociocultural Studies in Education at the University of Oregon, and I am continuing to pursue my interests in social justice and science education as an assistant professor at Moravian College in Bethlehem, PA.

Chapter 10
On the Possibility of Authorship in Science Education

Juliano Camillo

10.1 You Are Too Young for Theoretical Considerations! Who Are the Authors You Are Using to Support These Affirmations?

When the Portuguese arrived
In a heavy storm
He clothed the Indian
What a pity!
If it had been a sunny morning
The Indian would have undressed
The Portuguese
(Andrade 1954, p. 23)

It is not unusual to find narratives (or worldviews) wherein all that exists is split into different (or even opposing) realms. Examples of this are the separation between the divine and the human in many religions; between sublunary (ephemeral) and superlunary (essential) realms in the Aristotelian ontology; between the human and the natural (and subject and object) in modern science.

These splits become relevant, from the standpoint I am outlining here, to the extent that only very "special" groups of people (priests, philosophers, scientists, for example) are entitled to conceptualize the relations between separate worlds. They constitute the elites that are naturally gifted (and have "natural" permission) to "translate" the essence of the things, i.e. to access the true reality (which lies outside human activity) and produce valuable knowledge (communicating God's will to ordinary people; finding the place of things in the Cosmos; subjugating Nature to human desires, for instance).

J. Camillo (✉)
Federal University of Santa Catarina, Florianópolis, Brazil
e-mail: juliano.camillo@ufsc.br

J. Bazzul, C. Siry (eds.), *Critical Voices in Science Education Research*, Cultural Studies of Science Education 17,
https://doi.org/10.1007/978-3-319-99990-6_10

This resonates with decades (or even centuries) of science education practices based on preparing (or attempting to prepare) students for yet another level of scientific courses or, in other words, selecting talented minds to follow scientific careers. School science becomes an end in itself and exclusively centered on the canonical tenets of science (synthesized in classic manuals and textbooks). The better the science education, the better students approximate themselves to the "right answers" provided by science. At the end of this process, some enlightened people will officially be proclaimed an author(ity) in natural issues and explain/translate the way reality operates.

This becomes even more problematic when it is assumed that science education is the *sine qua non* for individuals to act as "reflective citizens" (OECD 2016) in a society increasingly permeated by scientific and technological issues. Democracy becomes directly associated with a specific form of knowledge (manifested, for instance, in international standardized tests that go over many local idiosyncrasies and without effectively taking into account what teachers, students, families, communities consider relevant as an educational project). In addition, this *modus operandi* reinforces the supposed neutrality and detachment of the science, taking it as absolute and ahistorical truth and proclaiming the division between those who produce knowledge and those who should consume it; between those who know and those who do not know.

From the moment I started my research in Science Education I was promptly located within the group of those who are meant to exclusively consume theories and methodologies of other researchers. Allegedly, I am not an author(ity), which would be evidenced by the lack of publication in high impact journals, and by not being part of a "high-level" international research community. For that, my work should follow the "natural path" of collecting empirical data to support specific theoretical approaches, instead of being allowed to develop autonomous thinking, judging, criticizing, evaluating, i.e. taking concrete activities (which would include others' theories and methodologies) as the fabric of my reflections. In this sense, the hierarchy and the concept of truth have been established based on "who is the researcher" criteria, and an epistemological-ontological split between reality and knowledge about reality (and between theory and empirical data) has been institutionalized in science education research.

I understand that this conservatism of science education (Lemke 2011; Burke and Bazzul 2016) cannot be grasped only in terms of a researcher's disposition/openness to dialogue (although these barriers do exist, including problems with language/communication) or simply in terms of attempts to maintain privileged positions (although this certainly is happening). It should be conceptualized in terms of hegemony, which "acts to 'saturate' our very consciousness, so that the educational, economic and social world we see and interact with, and the commonsense interpretations we put on it, becomes the world *tout court*, the only world" (Apple 1990, p. 5). This hegemony is not simply abstract, but "refers to an organized assemblage of meanings and practices, the central, effective and dominant system of meanings, values, and actions which are lived" (p. 5).

In this sense, we face the challenge of entering hegemonic spheres to push their boundaries and create (and participate in) new spaces with new ideas/researches/practices, giving new meanings to words and practices, seeking the emergence of new forms of activities, i.e. accomplishing the "untested feasibility" (Freire 2005). Here I choose to enter science education advocating for a critical philosophy of science education, which means to push some boundaries of this field to introduce new objects of reflection (Rodrigues et al. 2014), and to advocate for a specific meaning for "critical" that is not present in the mainstream research.

10.2 This Is Exactly What We Have Been Doing in Science Education...

Tell me, is the rose naked
or is that her only dress?
(Neruda 1991, p. 3)

It is not a huge challenge to advocate that concepts may have different meanings in different contexts. This polysemy takes place not only in everyday life, but also within groups that are not very open to multiple interpretations: think, for instance, how scientific concepts have evolved throughout the history, despite the attempts to control them. Disputes arise and this shows that concepts are not empty, but rooted in the complex net of human practices (Bakhtin 1981; Voloshinov 2000).

Nonetheless, special attention is required to situations wherein a false (or fragile) consensus is established, since people assume they completely share the same meanings or, in a more delicate way, take for granted that no other meaning is possible. Although there are numerous examples of false-consensus in science education (democracy, citizenship, scientific knowledge, scientific literacy; just to cite some of them), I want to bring something I experienced during the II World Conference on Physics Education in 2016, which took place in Brazil while the country was facing political turmoil.

It is worth noting that the political crisis (or generally speaking anything of the outside world) never entered the core discussions of the conference. Despite the fact that the conference theme was "Contemporary Science Education and Challenges in the Present Society: Perspectives in Physics Teaching and Learning", the word challenges largely referred exclusively to difficulties that teachers and students face when teaching and learning Physics the "way it is" – in its canonical form. Cultural, social, political, race and gender aspects, for example, might be visible in some cases, but only the extent to which they facilitate teaching and learning processes.

In a rare exception, researchers started a broad discussion about the meaning of science and its relevance to people's lives. A senior researcher took the floor to disapprove the path the discussion had taken, arguing that researchers in science education should focus on what they are meant to investigate, with precision and control: the microcosm of students' learning, leaving aside those imprecise and excessively broad aspects.

This modest example allows me to emphasize how explicitly power can be exercised in selecting what should be investigated, creating boundaries to the field and imposing consensus around the meaning of science education research.

In addition to the false consensus on what scientific research means, I also want to explore the notion of "critical", which is linked to my work. In many discussions, my use of critical was reframed as critical thinking, which stands for the idea that scientific knowledge has an intrinsic potential to promote skills and particular forms of reasoning that can be applied (and are desired) in various contexts. In other situations, critical was simply taken as opposite to naïve. From these perspectives, my work does not bring anything new to the field, since any good research is critical.

If we look carefully, critical is narrowly conceptualized most of the time, in terms of cognitive processes or skills that can be improved through practice (Bailin 2002). The realm of brain processes (or their correspondent observable behaviors) admittedly becomes the locus of interest for science education researchers, which is in tune with a specific conception of knowledge production and objectivity: neuronal activities can be measured with the necessary "scientific" precision. Not surprisingly, neuroscience has been expected to be (and is funded as) the new educational panacea, the way behavioral and cognitive psychology were.

Schulz (2009, p. 226) argues that "science education research continues to be heavily influenced [...] by theories coming from outside the discipline itself". Despite not being completely in line with the perspective presented by him, I do agree that we need "to develop an explicit *philosophy of science education*" (2009, p. 226), in order to "better attend to the needs of science teaching, learning and curriculum unique to it, and which (as an educational sub-discipline) it more properly shares with others in educational studies" (2009, p. 226).

Moreover, by reclaiming a notion of "critical" together with "philosophy", rather than engaging myself with another false consensus, I want to draw attention to the necessity and possibility of questioning the way knowledge is produced and circulated in the society, and make explicit "the underlying epistemological and ideological assumptions that are made about what counts as 'official' or legitimate knowledge and who holds it" (Apple et al. 2009, p. 3).

Critical, within this perspective, instead of a final stage or an abstract reference, has much more to do with an everlasting condition of making explicit *ideo-methodological* commitments of activities wherein we continuously engage, and grasp possibilities for social justice and transformation of reality. Here, Brecht (1992, p. 111) would warn us to observe the conduct of people:

> Find it estranging even if not very strange
> Hard to explain even if it is the custom
> Hard to understand even if it is the rule
> Observe the smallest action, seeming simple,

With mistrust
Inquire if a thing be necessary
Especially if it is common
We particularly ask you –
When a thing continually occurs –
Not on that account to find it natural
Let nothing be called natural.

10.3 This Only Makes Sense Within Your Framework

Look; the most important and nicest thing in the world is this: that people aren't always the same, they are not all of a piece and finished but keep on changing. They are in tune or out of tune. This is a great truth. It is what life has taught me.
(Rosa 1963, p.17)

I ask the reader to indulge in a little remembrance while I move forward with my discussion. This remembrance is about my mother's famous potato salad. Despite its renown, it was something I used not to like when I was a teenager. However, from the moment I started to like it, every time she cooked her famous dish, the conversation at the table was about how strange it was that I have changed my taste. I might say that from my mother's standpoint some changes are not hard to understand: I am getting old (which is natural) and now I am becoming gray (what was supposed to happen, since my father's whole family is). But how could it be possible that I have changed my taste (such an essential aspect)?

The point here is not potato salad. Rather, it is about a very recurrent and strong conception that lies behind my mother's expectation: what might be taken as natural; those inborn features that constitutes what I really am from the very beginning and should stay immutable throughout my whole life – my essence. The issue of what is natural/essential becomes imperative when, for example, the president of the one of the most prestigious universities in the world argues that "men outperform women in maths and sciences because of biological difference, and discrimination is no longer a career barrier for female academics" (Goldemberg 2005), which leads to a drastic simplification of such a complex problem, as if all gender inequality in science (and in the society likewise) had its roots in in supposed inborn differences, disregarding all the intricate (social, cultural, historical, economical, etc.) aspects that shape it. Moreover, Henrich et al. (2010) point out that what we largely consider typically human, related to fundamental aspects of psychology such motivation and behavior, for example, has been generalized from the least representative group of people: "WEIRD" (which stands for those who come from western, educated, industrialized, rich, and democratic societies). They argue that "there are no obvious a priori grounds for claiming that a particular behavioral phenomenon is universal based on sampling from a single subpopulation" (p.1).

The words above work as a preamble to something I have started outlining: the necessity of a critical perspective on human development in science education, at ontological and epistemological levels – radical, to use Marx's (2009) expression, standing for the idea of grasping things by their roots. Theories of human development are not merely abstract constructions, but they "are always intimately related, in a bi-directional way, to ideologies and policies of research and practice and have immediate practical ramifications in real life, worldly contexts, and everyday matters" (Stetsenko 2008, p. 473).

Radical perspectives are urgently needed to provide concrete alternatives for what means to be human and the role of science (understanding its limits and glimpsing new forms of doing it) in the process of becoming human. By failing to provide such alternatives, important educational aspects (such as motivation, intelligence, agency, and others) remain reductionist, i.e. converted to functions of the brain or other "internal" elements that have their own laws of development and are independent of what is going on outside the body.

Over several years at the university as a Physics student, I was trained to pursue a world without humans. The better (or more objective) the scientific research, the more effective was the eradication of human subjectivity of the achieved knowledge. In this perspective, no matter who the researchers are, no matter the society in which their research take place, they should be able to achieve the same results once things are what they are, and the laws of nature are there to be discovered.

Armed with that conception (and with a bunch of equations/theories/concepts, and well prepared to apply them to some everyday situation) I entered the classroom and became a Physics teacher. It was difficult for me to understand why students were not able to directly perceive all the things I was pointing out to them: electrons, electric current, electric potential difference, and many other entities that were "just there", being revealed by a multimeter, for instance.

It took a long time for me to deconstruct the idea of such a "self-evident" thing out there available to all students; something that could be perceived by everyone independently of their individual trajectories – their individual histories and particular cultures. It took a long time for me to realize that the specific form of knowledge I was presenting to them was forged under very specific cultural and historical conditions that are not immediately present in everyone's lives. In this sense, is there a place for human individuality in science education? Is there a real place for the individual trajectories and personal knowledge not only to the extent that they facilitate the transition to scientific knowledge during instruction?

In order to start building answers to these questions, my choice here is to explore the historical dimension of human development (Ilyenkov 2009; Leontyev 2009; Camillo and Mattos 2014), through a perspective that can be called *Potential Activity*. History in this context is not merely related to changes that might occur over time. On the contrary, it stands for the essentially unfinished nature of human beings, the process of becoming (Freire 2005; McLaren 2009). Human beings are always unfinished and they can be aware of such incompletion, which opens up the possibilities to overcome the idea that humanity has achieved some kind of "most developed stage" (due to its natural law of development), and also the idea that no other forms of human life can be built.

In this sense, not only do the past and the present take part in human development, but also the future: human consciousness/activity is future-oriented (Leontyev 2009). From the very beginning of their existence, individuals have to face the world "as problem to be solved" (Leontyev 2009), which means that they engage in/with the world, pursuing objectives, in/through a complex net of activities, whereby consciousness can emerge – not as a causal-temporal relation, but as a unity: activity-consciousness. Within this framework, human development is conceptualized in a concrete and historical perspective, without postulating any ontological break (or dualism) between human activity (what includes knowing, transforming, and becoming human) and an outside reality (whose core is essentially ahistorical) (Stetsenko 2008).

The problematic nature of reality (the complex net of activities that we can call "Problem-in-itself") can be grasped by individuals, making over a Problem-in-itself into Problem-to-itself. This means that human beings have the possibility to consciously build new forms of activities/consciousness, i.e. human individuality is not merely contingent, but also a project (Camillo 2015).

10.4 What Is This For? How Do You Apply This?

> *Somewhere there must be a garbage dump where the explanations are piled up. There is only one worrying thing about this panorama: what will happen the day someone can also explain the garbage dump.*(Cortázar 1984, p. 45)

It seems that any attempt to become an author in the science education field should pass this crucial experiment: the immediate applicability of the theory/reflection produced. Rather than fighting against the traditional question "Can you give us some practical example so we can see how it is applicable?", I will always engage with it, bringing back issues I have raised, thereby "applying" ideas!

Considering the complexity of educational phenomena, it is understandable that many times people expect to have theories/methodologies/gadgets that could be easily applied to solve a bunch of problems in different situations. Nevertheless, I consider relevant to move forward and explore the idea that such immediacy might be related to the way knowledge is conceived: as an increasingly faithful copy of an immutable reality. This brings back the problematic notion of "a" scientific method.

One might say that this issue has been overcome, and that nobody nowadays would defend some sort of scientific method in the way it has been caricatured by its critics. Nonetheless, supposedly overcoming the idea of a pictorial (positivist) scientific method does not mean that many of the underlying principles of Positivism have been overcome (Kincheloe and Tobin 2009).

Two examples might be interesting here. The first one happened when I was talking to a colleague about some ideas I am raising here. He affirmed that as a physicist he could not live without the conviction that the reality is out there and our knowledge has been progressively approximating to it. The second is about an encounter I had with an Italian physicist who was spending some time in Brazil. He had in mind a project to help traditional communities of fishermen to build better boats using Physics. According to him, those fishermen have been building boats 10% less efficient than they could be.

Beyond the split between subject and object, and the commitment to an immutable sphere of the reality, where the truth professedly lies (expressed in the first example), the second one makes explicit the certitude that scientific knowledge is abstract and universal, i.e. the knowledge that it is in itself, by its own nature, higher than all other forms regardless of the context. In fact, scientific knowledge would be able to improve boats' efficiency. But to what extent, by narrowly assuming Physics' standpoint, this "improvement" would be done taking into account people's real life and their real needs, or questioning, for instance, what 10% efficiency means in face of this "cultural invasion"?

Standard narratives about science tend to avoid the ideological dimension and do not (many times deliberately) grasp how methodological and ideological (or I would say ideo-methodo-logical) dimensions are entangled. Despite the fact that scientists often make use of epistemologies that would not fit into what counts as science, this does not appear in these standard narratives and they are often not officially inserted into the structure of the scientific enterprise. Feyerabend (1975), for example, has shown that Galileo was not loyal to a methodological monism when defending heliocentric model, which was essential to the development of the science. Others like Fleck (1979), Latour and Woolgar (1992) have shown that what happens outside of scientific community is much more present inside the scientific enterprise than its widespread image of science has been able to represent.

By emphasizing that "the outside world" takes part in the scientific research, I do not want just to advocate that scientific agenda is biased by social interests (which in fact happens in many situations), keeping untouched the way knowledge is produced; i.e. *what* is to be investigated is influenced by the society, but *how* it is investigated does not suffer any effect. In the sense, the reality is out there and, at some point in the history, if there is the interest of some abstract scientist/society, knowledge will be invariably produced. I also do not want either to advocate that science can be converted into pure discursive consensus among those who are involved in its production, overstepping any particularity of scientific knowledge.

Taking the concrete and historical perspective I am outlining here (Potential Activity), the criteria of truth should not be only methodological (as a procedure that guarantee reliability for the process of knowing something in general), but ontological; i.e. rooted on the very nature of the reality (as the totality of the historical process), and for that, it is unequivocally situated (historically).

Let me take as an example one scientific notion to illustrate how human development is an integral process by which humans make up themselves transforming the reality, i.e. through/and in the reality actively. Within this perspective, the atom is not an attempt to approximate more and more to a static entity (reality in itself), but it is, on the contrary, the very expression of one facet of human development, in a concrete historical stage, entangled with the totality of the (processual) reality. Paradoxically, any attempt to "find" the real nature of the atom (without the presence of human being) brings us to rely increasingly on the most developed apparatus, emerged from the very human activity (for example, LHC/CERN).

In putting the question about human development, I think it is possible (and necessary) to overcome the conception in which reality is constituted by fixed things, with their intrinsic properties, toward a conception of things as endless processes. In this sense, "atom" express the totality of complex relations that are established around them and synthesized through them, with no dualism between some sort of "atom in itself" and facets we know about it. Knowledge turns out to be the appropriation of the concrete determinations that make possible to operate with atoms to transform the reality in concrete situations, not only methodologically but ideo-methodo-logically, i.e. taking into account the totality of real life.

Every time that atom (as a totality of relations) is used (i.e. that some transformation is operated through it), reality is not only transformed, but also the notion and materiality of atom, since it has to continuously face a new and different reality. Human subjectivity (that takes part in this complex process of the transformation of the reality) evolves through time – and this is a key relation between the development of science as human development.

I want to end with the example of the movement of Brazilian students, that occupied hundreds of schools in 2016 in order to take education into their own hands. They created an organization that allowed them to take care of the school in its wholeness (cleaning, cooking, deliberating, taking decisions about what to learn). This movement happened not only without the State, but against the State and its repressive forces materialized in the violence used by the military police. This movement clearly shows us the depletion of an educational model, and a model of human being, in which students have no agency and no voice in building their own lives through education.

> I understand your point of view, but there are too many things that are open there. The object of your investigation is excessively wide. I am going to recommend you an author to help you…

References

Andrade, O. (1954). The mistake of Portuguese. In L. S. Downes (Ed.), *An introduction to modern Brazilian poetry* (p. 23). São Paulo: Clube da Poesia do Brasil.

Apple, M. W. (1990). *Ideology and curriculum*. New York: Routledge.

Apple, M. W., Au, W., & Gaudin, L. A. (2009). Mapping critical education. In M. W. Apple, W. Au, & L. A. Gaudin (Eds.), *The Routledge international handbook of critical education* (pp. 3–19). New York: Routledge, T&F Group.

Bailin, S. (2002). Critical thinking and science education. *Science & Education, 11*(4), 361–375. https://doi.org/10.1023/A:1016042608621.

Bakhtin, M. (1981). *The dialogic imagination*. Four essays (C. Emerson & M. Holquist: Trans.; M. Holquist, ed.). Austin: University of Texas Press.

Brecht, B. (1992). *The Jewish wife and other short plays*. New York: Grove Press.

Burke, L. E. C.-A., & Bazzul, J. (2016). Locating a space of criticality as new scholars in science education. *Cultural Studies of Science Education*, 1–15. doi:https://doi.org/10.1007/s11422-016-9735-7.

Camillo, J. (2015). *Contribuições iniciais para uma filosofia da educação em ciências*. Doctoral thesis, Ensino de Ciências (Física, Química e Biologia), University of São Paulo, São Paulo. doi:https://doi.org/10.11606/T.81.2015.tde-25112015-144311.

Camillo, J., & Mattos, C. (2014). Making explicit some tensions in educational practice: Science education in focus. *Cultural-Historical Psychology, 10*(2), 110–115.

Cortázar, J. (1984). *A certain Lucas*. New York: Knopf.

Feyerabend, P. (1975). *Against method: Outline of an anarchistic theory of knowledge*. London: NLB.

Fleck, L. (1979). *Genesis and development of a scientific fact*. Chicago: University of Chicago Press.

Freire, P. (2005). *Pedagogy of the oppressed*. New York: Continuum.

Goldemberg, S. (2005). Why women are poor at science, by Harvard president. *The Guardian*. Retrieved from https://www.theguardian.com/science/2005/jan/18/educationsgendergap.genderissues

Henrich, J., Heine, S. J., & Norenzayan, A. (2010). The weirdest people in the world? *Behavioral and Brain Sciences, 33*, 61–135.

Ilyenkov, E. V. (2009). *The ideal in human activity*. Marxists Internet Archive. Pacifica.

Kincheloe, J. L., & Tobin, K. (2009). The much exaggerated death of positivism. *Cultural Studies of Science Education, 4*(3), 513–528. https://doi.org/10.1007/s11422-009-9178-5.

Latour, B., & Woolgar, S. (1992). *Laboratory life: The construction of scientific facts*. Princeton: Princeton University Press.

Lemke, J. (2011). The secret identity of science education: Masculine and politically conservative? *Cultural Studies of Science Education, 6*(2), 287–292. https://doi.org/10.1007/s11422-011-9326-6.

Leontyev, A. N. (2009). *The development of mind*. Pacifijica: Marxists Internet Archive Publications.

Marx, K. (2009). *Introduction to a contribution to the critique of Hegel's philosophy of right*. Marxists Internet Archive. https://www.marxists.org/archive/marx/works/1843/critique-hpr/intro.htm

McLaren, P. (2009). Critical pedagogy: A look at the major concepts. In A. Darder, M. Baltodano, & R. Torres (Eds.), *Critical pedagogy reader* (2nd ed., pp. 61–83). New York: Routledge.

Neruda, P. (1991). *The book of questions*. Port Townsend: Copper Canyon Press.

OECD. (2016). *PISA 2015 assessment and analytical framework: Science, reading, mathematic and financial literacy, PISA*. Paris: OECD Publishing. https://doi.org/10.1787/9789264255425-en.

Rodrigues, A., Camillo, J., & Mattos, C. (2014). Quasi-appropriation of dialectical materialism: A critical reading of Marxism in Vygotskian approaches to cultural studies in science education. *Cultural Studies of Science Education, 9*(3), 583–589. https://doi.org/10.1007/s11422-014-9570-7ID.

Rosa, J. G. (1963). *The devil to pay in the backlands*. New York: A.A. Knopf.
Schulz, R. M. (2009). Reforming science education: Part I. The search for a philosophy of science education. *Science Education, 18*, 225–249. https://doi.org/10.1007/s11191-008-9167-1.
Stetsenko, A. (2008). From relational ontology to transformative activist stance: Expanding Vygotsky's (CHAT) project. *Cultural Studies of Science Education, 32*, 471–491. https://doi.org/10.1007/s11422-008-9111-3.
Voloshinov, V. N. (2000). *Marxism and the philosophy of language*. Cambridge, MA: Harvard University Press.

Juliano Camillo taught Physics in elementary schools for many years and nowadays is Professor of the Department of Teaching Methodology and of the Graduate Program in Scientific and Technological Education at the Federal University of Santa Catarina in Brazil. He is interested in developing a philosophy of science education, concentrating his efforts in understanding the relations between human development and science education.

Chapter 11
Beyond Levinas' Other: My Journey Reimaging Science Education

David Blades

When a paper appears in a publication, for example as a journal article or chapter in an edited collection, authors rarely provide context that might inform readers of the author's journey with this paper from conception to publication. Sometimes this journey is easy, especially when reporting data from a particular study; but I have found that some papers and projects seem to develop a life of their own, a writing process that can take an author to unanticipated ideas that sometimes have uncomfortable conclusions. An example is also my favourite publication, *Levinas and an Ethics for Science Education* (Blades 2006). This chapter is a narrative of how this paper developed. I share this story in the hope of revealing the considerable struggle embedded in some, if not most, academic writing. My intention in sharing the narrative that follows is animated by the desire to reveal how my individual self-as-author of this paper is a suspect presentation. In his essay, "What is an Author?" (1984) Michel Foucault points out how some texts historically become foundational to fields of discursivity; he uses Sigmund Freud's work and others to illustrate his point, noting that, "in this sense, they are very different, for example, from a novelist, who is, in fact, nothing more than the author of his [sic] own text" (p. 114). But in making this distinction, Foucault misses another possibility, namely that a text may not be the sole creation of an author but a product itself of what I will call "discursive injections" throughout the writing of the work. Some, if not all, academic papers may go on to become foundational themselves but they first emerge as a result of voices, discourses, and historical timing that led to the creation of this publication, a work that appears to be the creation of a single author that is actually the result of a momentary crystallization of many discursive events and voices. Knowing that some papers (if not most) are the results of many voices and events invites us as readers to ask questions as we read, which can move reading from the

D. Blades (✉)
University of Victoria, Victoria, Canada
e-mail: dblades@uvic.ca

J. Bazzul, C. Siry (eds.), *Critical Voices in Science Education Research*,
Cultural Studies of Science Education 17,
https://doi.org/10.1007/978-3-319-99990-6_11

assumption of meaning to a conversation with the text that questions meanings I discover, since behind what is read are stories about what led to the appearance of the text. This conversation in turn can become a humanizing of a publication in the way the text is actually a crystallization of moments in an author's life: We are invited to learn more about the author and in that invitation the writer become less anonymous and so writing becomes less mysterious and more an opportunity for all of us with stories to tell.

11.1 Interjections

Levinas and an Ethics for Science Education was published in 2006 in *Educational Philosophy and Theory* (Volume 38, #5) as part of a special issue on the philosophy of science education. The editor of the special issue (and who contributed a paper as well), Michalinos Zembylas (2006), wrote in his introduction that, "this special issue provides an overview of the state-of-the-art developments in the field of science education by drawing upon advanced contemporary philosophical thinking from scholars in the field" (p. 585). I almost missed the opportunity to contribute to this collection. My initial submission was a somewhat hesitant exploration of how philosopher Emmanuel Levinas' radical view of the Other might inform a possible ethical foundation for science education curriculum. But this first paper remained, for the most part, in what was for me the safe territory of theoretical speculation: Explain to readers Levinas' approach to solving a key issue in the field of ethics and then hypothesize what this solution might mean for science education. The editor's response to this first draft was as insightful as it was brief; Zembylas complimented the topic and direction of the paper, but felt that I was "holding back" on something important, something I needed to say but that was not forthcoming in the draft. Sometimes when we write we choose the safer route, when a more critical and even radical idea lurks underneath our writing. Academic writing can be seen as an opportunity and even responsibility to bring forward these ideas into public discourse if we are serious about invoking conversations about present practices and possibilities for change.

While disappointed by his response, his assessment was profoundly accurate: I *was* holding back on a full exploration of Levinas' ideas. The first submission mostly restated previous arguments I had developed about the need for a post-structural approach to science education curriculum, with Levinas' ideas appearing as an alternative possibility, but not explored thoroughly. I had built the paper around an experience I had as a high school science teacher when we dissected frogs. As I wrote about this dissection it occurred to me that Levinas' notion of the Other could be extended to other living things, such as frogs. Did I have an ethical responsibility to the frogs brought into my classroom? This question seemed to me to be an under-examined aspect of what Levinas is talking about.

I considered dropping out of the special issue, but for another reason. During the time of this paper I was also serving my Faculty as Associate Dean of Teacher

Education. This administrative responsibility was entirely consuming and left little time for thought, reading or reflection, all needed to rewrite the paper. I honestly didn't think that I would have the time or space to revise the paper in ways Zembylas and I needed. But, during this consideration I was surprised by a "discursive injection" in the form of an encounter with a student in our teacher education program. The teacher candidate had failed two student-teaching experiences in schools and now had to meet the Associate Dean—me. As we chatted in my office, suddenly the words of Levinas came to me—his call to attend to the Other, to *really* listen to the Other and to put aside all anticipations about the person. Levinas makes a distinction between what he calls the "said"—the existing dialogue, narrative, or discourse about something or someone—and the "saying"—the unique voice of appearance of the Other in a particular encounter (Levinas 2002). Usually, Levinas argues, the said overshadows the saying, preventing us from hearing the voice and needs of the Other. I decided to focus on the teacher candidate, to their saying and to just listen. What emerged was sad, for both of us; the candidate admitted to being forced into teaching by supportive parents, but that this was not really a career choice the candidate really wanted. I listened more and, slowly, and with tears, a life story emerged and in it, some possible other career directions; mostly, though, I just was present for this person as fully as possible. The meeting ended well, I believe, and the candidate, who decided to withdraw from the program, seemed genuinely relieved and happy.

This encounter encouraged me to return to Levinas' writings and to consider in more practical terms the implication for science education. I continued to re-read and reconsider my first submission. But I was still frustrated finding time to actually work on this paper due to the demands of administration. Finally my assistant, who controlled my schedule, suggesting setting a block of time each week entirely devoted to working on this paper. Protecting this space was difficult and I was grateful to have an experienced assistant who could move meetings, shift around appointments and generally protect my "research time" so that I would write; without this interjection into the busy life of an administrator, I believe that this paper would never have been completed. The second paper was substantially better and became the published form, a version only possible because of the three people that made the rewrite possible: the editor's insight, the student encounter and my assistant's skill with my schedule.

The paper on Levinas and science education traces back to an issue arising from the final chapter of my dissertation. My doctoral research was a case study of an attempt by a provincial government in Canada to change the province's secondary school science education towards a more Science-Technology-Society-Environment (STSE) emphasis. An STSE approach was advocated and articulated in the Canadian Council of Ministers of Education's publication *Common Framework for Science Education* (1997). The province I studied was one of the first in Canada to adopt an STSE approach to science education, but the process of adoption collapsed as the science education curriculum folded back to a resemblance of the status quo. The question I examined seemed initially simple: Why did this happen? My studies of attempts at curriculum change were not illuminating. Most publications at the time considered curriculum change a technical challenge, a belief that having all the factors line up would guarantee change. An example of a popular work advocating such technical-rational approaches was Michael Fullan's *Meaning of Educational Change* (1983). My review of this work and others led to the surprising conclusion that, in the province I studied, those working on the change *had* considered all the factors and *still* the change did not occur. I therefore sought other explanations of change attempts, which led me to James Marshall's 1989 article on the implications of Foucault's concept of "power" to education—one of the first in education to examine Foucault's ideas. I found the idea of power compelling and useful in explaining curriculum change as a discursive practice, bounded and affected by what I called "procedures" of power. That dissertation, which later became a book, was one of the first "post-structural" approaches to understanding the dynamics of curriculum change (Blades 1997).

The final chapter of my book left a lingering question: What possibilities exist for ethics after post-structural critique and under the general umbrella of post-modernism? Foucault suggests a possibility: He argued that, "the real political task in a society such as ours is to criticize the working of institutions which appear to be both neutral and independent; to criticize them in such a manner that the political violence which has always exercised itself obscurely through them will be unmasked, so that one can fight them" (quoted in Rabinow 1984, p. 6). In their work, *Ethics and the Foundations of Education* (2003), Patrick Slattery and Dana Rapp provide examples of Foucault's call to critique, considering their personal examples of resistance to the defining and sometimes violent discourses of education as ethics-though-action. Still, this response does not clarify a philosophical basis that justifies this action, reasons why one should act in certain ways; this question is especially difficult in the light of a post-modern rejection of any universal, foundationalist approaches, including a system of ethics (Lyotard 1989). John Caputo (1993) argues a way out of this difficulty in his work, *Against Ethics*, where he suggests our actions can be guided by a non-foundational sense of *obligation*. In this work he draws inspiration from the ideas of Emmanuel Levinas, although finds Levinas' approach to ethics "too pious" (p. 15). All of these authors—and there are more I could list as well—represented discursive injections to the lingering question of ethics arising from my dissertation research, but Caputo's work introduced me to Levinas and my thinking about the question of ethics and post-modernity took another, radical turn.

11.2 Levinas' Answer to the Key Issue in Ethics

Anyone studying ethics will encounter the ancient dilemma of applied ethics that has challenged philosophers for centuries. In order for ethics to be concerned with social justice (and not just inward character, for example), ethics must consider interaction with others, but how can one know the Other? The dilemma is that if I assume the Other is somehow like me in some sort of transcendent way, then I have subsumed otherness with the Same, effectively denying the possibility of otherness, or alterity.

My thinking about this question led me to study phenomenology in more depth and especially Martin Heidegger's *Being and Time* (1926/1962). One of my friends, Gidi Nashon, was involved at this time in an extensive review of Heidegger's works and life; Gidi and I talked often about the difficulty reconciling Heidegger's concepts of Being with the events of Heidegger's life and neither Gidi or I felt that Heidegger's views provided a sufficient basis for an ethical system of social justice. But Gidi shared with me a little book by one of Heidegger's students, Levinas; *another* voice pointing me towards the work of Levinas! In *On Escape*, Levinas (2003) argues that Heidegger's notion of transcendence of Being through our common destiny of death still subsumes the Other to the Same; a similar issue exists with Heidegger's later work on humankind as technological beings.

In reflection, I realize that it was inevitable that I would encounter Levinas' work, since he effectively solves the question of the Other. Levinas managed to find a solution by adopting the radical and somewhat surprising view that we can *never* know the Other, that the Other is infinitely and always a stranger to us. Though what amounts to as a rejection of transcendence as a basis of ethics, Levinas instead argues that ethics begins with the acceptance of the infinite "alterity of the Other" and that ethics begins with listening to the call of the needs of the Other, without any expectation of reciprocity.

Right at the time I was trying to understand Levinas' work, *Otherwise Than Being* (2002), I was contacted by Zembylas with an invitation to submit a paper to the special issue of *Educational Philosophy and Theory*. I rather naively thought that a paper on Levinas could be a useful addition to the collection and sent him a proposal. Of course, sometimes an accepted proposal becomes a significant challenge! All the voices of scholars I knew and those I've never met in person seemed to crystalize in the form of an opportunity to write this paper, but everything ground to a halt as I increasingly realized that Levinas' "solution" to the question of the Other would be very difficult to apply to science education.

11.3 A Frog and the Call of the Non-human Other in Science Education

Levinas' work provides a clear and brilliant solution to the philosophical question about the alterity of the Other and his work also reveals ways we might be and become with each other. I could see how this might work out in practice and Levinas also gives us some advice in interviews on learning to attend to the Other. In science education, hearing the call of the Other could be conceptualized as a form of interaction in classroom dynamics, but I struggled to make sense of Levinas' ideas for the *science* in science education. Pondering this question in my university office way day, I was interrupted by a dead frog.

I'm not sure why I placed a photograph of a frog about to be dissected (ventral side up) on the bulletin board above my desk, perhaps because the photograph marked a key turning point in my career as a secondary school teacher. I used to dissect frogs with my students as a way of helping them realize the physical structure and general anatomy of a digestion system—and the activity used to be prescribed for secondary school science education. However, I also phased out this activity and never really reflected on why. As I gazed at the photo, a question formed: Had this frog, this particular frog dead, lying in a dissection tray, *also been an Other*?

The injection of this past memory with a new question was disturbing. Now, it might seem obvious that Levinas' phenomenology of the Other should naturally extend beyond the human Other, but Levinas actually never extrapolates his discussions of ethics beyond inter-human relationships. In fact, Levinas was *opposed* to considering the Other anything else than a human Other. I could see his point, though—while our daily encounters with pets and other animals might lead us to consider these companions as Other, why stop with these life forms? Where does one draw the line on considering obligation to the Other's needs, which Levinas considered the very foundation of a philosophy of ethics?

But should we also not consider that other animals, such as frogs, *have needs* and therefore also calls out, obligating us? And would not this call, which Levinas identifies as "responsibility," have implications for science education? These questions, which I had not really expected when I agreed to take on the paper, came to dominate my thinking. I decided to research frogs and the particular species we used to dissect, *Rana pipens*. My studies revealed a decline in the population of these frogs and actually *all* frogs worldwide. As well, I discovered further information about the sensitivity of frogs, especially their skin, to environmental pollutants. The discoveries about skin sensitivity led me to a more thorough study of frogs themselves: their lives, complex ecological relationships and amazing adaptations. I became very vocal about frogs (to the dismay of friends and family) and through my study of these amphibians I came to a disturbing question: What gave me the right to kill these animals for dissection? Now of course there is a huge body of literature about human relationships to animals and very strong critiques of our collective approach to animals as a resource to be used any way we believe benefits humankind, but somehow this discussion was never really part of my pedagogy as a science teacher—why was this?

Draft of the paper on Levinas increasingly took up such questions and in working though these a sense emerged on how Levinas might inform science education curriculum. I used the popular "STSE" acronym (and approach to science education that emphasized the connections and interplay between science, technology, society and the environment) as a way to reconceptualise science education curriculum development, finally arriving at the fairly uncontroversial and not-very-original argument that STSE should be refocused with the "E" at the centre of all curriculum development and the other perspectives around this core, an "E-STS" approach. I then sent the first draft of the paper off to the Editor.

He was correct that I was holding back on the implications of Levinas' work—I still had not fully appreciated what I was missing. I decided to read commentaries about Levinas' work to see if anyone else took up the question of the non-human Other.

11.4 What Is Science?

I discovered that at the time almost no one had considered fully considered whether Levinas' approach to ethics could be applied to the non-human Other, although many noted the importance of such an inquiry. It was at this moment that I experienced the important discursive interjection in the evolution of my paper: John Llewelyn's book, *The Middle Voice of Ecological Conscious* (1991). After noting Levinas' argument that only that which can speak owes us a response, Llewelyn asks, "but can the Other who calls me to responsibility, whether direct or indirect, be a being that cannot speak?" (p. 194). He then unpacks the notion of "a being" to even include what we would describe as non-living; using the example of a hammer Llewelyn invites readers to consider our obligation to even this technology arguing that, "we should not to treat tools only as means" (p. 259).

I was completely shocked and astounded by his arguments, which present a post-Levinasian view of the Other as every-thing. To this day, Llewelyn's argument haunts my thinking and I consider his book, and this section in particular, one of the biggest influences on my thinking. I could see an immediate implication of this argument: If any-thing can be the Other and if the Other, as Levinas claims, is unique, then we have to face this question: "what is a category"? Are all distinctions arbitrary, even illusions? And where does this leave modern science, an activity founded and devoted to the labelling and grouping of every-thing?

I needed to address these questions and had a planned trip to the United Kingdom, so I found the email address for Dr. John Llewelyn at Edinburgh University and decided to see if he would have time to meet. I was delighted with an immediate invitation! So, I included in my UK trip a meeting with John in his home city of Edinburgh. I also discovered that he was a personal friend of Levinas and Jacques Derrida and the three of them also discussed the question of the non-human Other (Derrida's last book, *The Animal that Therefore I Am* (2008), takes up the question of the non-human Other in detail). How did I meet Llewelyn? I simply emailed him and asked to meet! This is another example of the discursive events behind a paper and the struggles unseen in the final, published version. At our meeting, John shared with me that he believed one should not contain the concept of the Other; I asked how far one could and should take this concept and he turned the question back to me. We discussed the idea of obligation towards a hammer and if a hammer has "needs"—and it seemed to us that it does, in that this technology, as something produced from iron ore and felled trees, *requires* us to look after it, to be thoughtful in using a hammer and to ensure that this technology is not ruined through careless neglect. I shared how such attitudes towards the world, both as it appears as "nature" and is made through technology, has parallels with some indigenous views of the world, and he agreed. I left our meeting inspired and challenged to take Levinas' original ideas to the world more fully and, I can assure you, I treat the hammer I have at home with a new respect and appreciation since that meeting.

A post-Levinasian approach to the Other means that science, as we know the word, would have to be radically reconsidered from a practice of conquest to a constant state of unknowing. This does not mean that we can't try to know the Other, only that we never can, fully. The strange implication of post-Levinasian approach is also that we can't even assume categories of things, such as "frogs" actually exist—each frog would be, from a post-Levinasian perspective, unique, individual and entirely the Other, a stranger to us. All the scientific "said" about frogs would thus become suspect or, better, suspended, as we approach *each* form as something new to us, something that has needs and therefore presents an obligation to us. And this act of approaching in newness, in listening, would be an ethical foundation for science education. This foundation also presents an opportunity to examine and include in science education the epistemologies of those who have long-standing traditions of viewing the non-human as Other, such as indigenous cultures. This exploration does not refuse categories, but argues for an approach to teaching about science where such labels and groupings from any culture, including the culture of science, should be held lightly in the recognition that they are, after all, useful fictions open to change.

This was troubling to me, since a post-Levinasian approach to science education would actually undermine much of what is assumed and presented as science education in today's schools, especially the classification of life in Biology or organization of types of molecules in chemistry. But I could see how science could also provide understanding on how to meet the needs, if I can use this phrase, of technologies and also create a more ecologically-responsible citizen. I worked out the ideas of this as best I could in the new draft of my paper and resubmitted, confident that this rethinking of the very foundation of science education was what the Editor saw lurking but unexpressed in my first draft; it was and the paper was published.

11.5 Informed Representation: Authorship as Crystallization

As Foucault argues in his essay, "what is an Author?" (1984), "even when an individual has been accepted as an author, we must still ask whether everything that he [sic] wrote, said, or left behind is part of his [sic] work" (p. 103). It is not that the author has disappeared, argues Foucault, or effaced in some way, but that the *idea* of authorship usually neglects to consider the discourses that inform, regulate and extend the discourse of authorship. This is why he asks, "What" is an author. To his post-structural inquiry we might also ask: who is the author of a work? In this paper, the question doubles, for the simple answer is that I am the author of both this work and the paper I have discussed. Yet I have shown that the ideas discussed in my paper on Levinas and science education developed from an infusion of the voices and ideas of others, notably Llewelyn, Caputo, Nashon, and of course Levinas, but also everyone who accompanied me on the journey that was writing this paper, which included the interjection of the voice of a frog and the hammer I use at home. Researchers often claim that we, like Isaac Newton, stand on the shoulders of those who went before us, but I believe that metaphor is too linear and hierarchical; to the question, "Who is an author?" I prefer the image of a researcher walking with linked arms with many, many people and many things (if even such a distinction can be made), an author who can only respond: "us." That includes the paper before you as well—there are struggles, voices, and interjections that also influenced the directions and ideas of this paper—that while it carries my name, is not solely authored by me and there is a story behind this paper, too.

I believe that bringing to light the journey and struggles of a paper is important work. Should every publication share, even briefly, the journey of its creation? Likely not, but we should approach each and every publication as a moment, a crystallization of events into words where there are many authors, each part of a complex, dynamic network of discourses that led to the moment of publication. Knowing this makes the words in front of any reader more than just text, more than a technology. Adopting this attitude helps us to counter forces that direct writing as a form of production to meet quotas for academic promotion: How many papers have you published? Instead we might humanize the discourse of academic meritocracy by sharing the stories of some of our publications and the struggles and joys of discovery that led to a published text. Sharing these stories, as I have with you (however briefly) is hopeful in the invitation to also share in the questions animating the writing: What do you think about this or that? Why is this an issue? And to this collection I might bring to the mix my specific question about science that continues to haunt me since writing the paper about Levinas: What *are* categories? Are categories of things what I am calling these days, "useful fictions" or is there something transcendent about a category—like, "frog" or, "author"? And if you have some ideas about these things, let's chat.

References

Blades, D. (1997). *Procedures of power & curriculum change*. New York: Peter Lang.
Blades, D. (2006). Levinas and an ethics for science education. *Educational Philosophy and Theory, 38*(5), 623–640.
Caputo, J. (1993). *Against ethics*. Bloomington: Indiana University Press.
Council of Ministers of Education, Canada. (1997). *Common framework of science learning outcomes K to 12*. Ottawa: Council of Ministers of Education, Canada.
Derrida, J. (2008). *The animal that therefore I am* (D. Wills, Trans., M.-L. Mallet, Ed). New York: Fordham University Press.
Foucault, M. (1984). What is an author? In *The foucault reader* (pp. 101–120) (P. Rabinow, Trans.). New York: Pantheon.
Fullan, M. (1983). *The meaning of educational change*. New York: Teachers College Press.
Heidegger, M. (1962). *Being and time* (J. Macquarrie & E. Robinson, Trans.). San Francisco: Harper. (Original work published 1926).
Levinas, E. (2002). *Otherwise than being* (A. Lingis, Trans.). Pittsburgh: Duquesne Press.
Levinas, E. (2003) *On escape* (B. Bergo, Trans.). Stanford: Stanford University Press.
Llewelyn, J. (1991). *The middle voice of ecological conscious*. London: Macmillan.
Lyotard, J.-F. (1989). *The postmodern condition: A report on knowledge* (G. Bennington and B. Massumi, Trans.). Minneapolis: University of Minnesota Press.
Marshall, J. D. (1989). Foucault and education. *Australian Journal of Education, 33*(2), 99–113.

Rabinow, P. (1984). Introduction. In P. Rabinow (Ed.), *The foucault reader* (pp. 3–29). New York: Pantheon.
Slattery, P., & Rapp, D. (2003). *Ethics and the foundations of education*. Boston: Allyn and Bacon.
Zembylas, M. (2006). Introduction. *Educational Philosophy and Theory, 38*(5), 585–587.

David Blades is Professor of Science Education and Curriculum Theory in the Faculty of Education at the University of Victoria and also Co-Director of the Centre for Excellence in Teaching and Understanding Science at the University of Victoria. He is won many awards for teaching, including a 3 M National Teaching Fellowship in 2016, he also holds a Ted Aoki Award for his contributions to the field of curriculum studies. In his spare time, which is never enough, David loves to explore with his family the bays and marinas of the Pacific Northwest aboard their boat, *Almirena*.

Chapter 12
Maintaining Our Critical Work: Stories of Curriculum Making in Initial Teacher Education

Jenny Martin and Donna King

In this chapter, we reflect upon the process of curriculum making in initial teacher education (ITE). Even though we work in different States of Australia, Donna at Queensland University and Jenny at the Australian Catholic University in Victoria, we draw parallels in our experiences of implementing critical approaches to science education with preservice teachers at our respective institutions. Critical approaches to science education for us question the purpose of science education and are transformative, affording students the opportunity to challenge accepted mainstream practices and ultimately empowering them to change how the world is enacted in classrooms and beyond. Like Claudia Mitchell, Sandra Weber and Reiko Yoshida, we take up the challenge of "teaching to change the world" (2008, p. 139) and hope that our initial teacher education students (preservice teachers) will also.

Our work in curriculum development in ITE is informed by Environmental Education for Sustainability (EEfS). EEfS is an educational approach that began in the 1990s to address mounting concern over the environment and development problems occurring in society. During this time, the environmental education consortia took a stance arguing for an educational approach that focussed on contemporary issues impacting the sustainability of our planet. This form of environmental education "is concerned with integration of the complementary disciplines of environmental and development education and requires reconciliation between environmental conservation and economic development" (Tilbury 1995, p. 197). As such, environmental education became focused on "improving the quality of life of all citizens under the focus of EEfS" (Tilbury 1995, p. 197). Such an education strategy was

J. Martin (✉)
Faculty of Education and Arts, Australian Catholic University, East Melbourne, Australia
e-mail: jenny.martin@acu.edu.au

D. King
Faculty of Education, Queensland University of Technology, Brisbane, Australia
e-mail: d.king@qut.edu.au

J. Bazzul, C. Siry (eds.), *Critical Voices in Science Education Research*,
Cultural Studies of Science Education 17,
https://doi.org/10.1007/978-3-319-99990-6_12

designed to change students' attitudes and practices to reflect an ethic for living for sustainability. This required students to explore links between their own lives and the broader environmental and development concerns (e.g., consumerism, global warming, and increasing world population). The education strategy underpinning EEfS continues today and aims to empower students to critically view science related to sustainability issues and to adopt practices that may transform their own lives.

Informed by sociocultural theory, any change in the enactment of the world requires collective action, which we believe needs to be ethically motivated towards social and ecological justice. Therefore, asking our preservice teachers to engage in action, including written and spoken action, towards transformative goals informs our curriculum making in initial teacher education. We illustrate this in the next section in our separate stories of curriculum making by elaborating assessment tasks we have each designed for our different teacher education courses and reflecting on their implementation in our institutional contexts.

12.1 Stories of Curriculum Making

In her story, Donna draws upon her email correspondence to preservice teachers regarding the integrated assessment task they were undertaking, emails from the academic staff from three discipline areas collaborating in its implementation as well as her own diary entries to tell her story of curriculum making. This task, situated in a course on EEfS, required students to use an inquiry-based approach to offer a solution to a sustainability issue in their local community by integrating knowledge they had learnt in the course across five Key Learning Areas: Science/Technology, Arts/Languages Other Than English (LOTE), Studies Of Society and Environment (SOSE). Through engaging in this transformative task, students were challenged to use their personal learning experiences to draw implications for their pedagogical practices as a teacher. The project was viewed as innovative in the course and in the institution and Donna received a teaching award from her institution, yet it was abandoned after its fourth iteration. Donna reflects upon her leadership in the project, its emotional demands and institutional practices that could have better supported the project.

Jenny frames her story by sharing an example of a preservice teacher's engagement in a course on EEfS using the Ethics of Care (see Carter et al. 2018). The preservice teacher is representative of many of the primary preservice teachers in the course who generally have had limited success in science previously. The task required preservice teachers to relate their learning in science to an improved capacity for caring in an online reflective journal or blog. Jenny suggests that although the preservice teacher disagrees with the scientific view of evolution, the journal entry demonstrates the engagement of a student who may have otherwise rejected science. Such success, she explains, is shared with her colleagues past and present, since the development of their courses is a collaborative and iterative process and one that has occurred over many years of scholarship (e.g. Martin and Carter 2015). Jenny reflects upon her emotional response to a course review that threatened the viability of their work and how intellectual freedom can never be a taken-for-granted given.

We employ the term *territory* in the title of our narratives. For us, this word highlights a relational ontology consistent with sociocultural theory (Stetsenko 2014). Social action can be understood as located between persons and in moments enacted jointly as a kind of emergent territory or space, either directly with other people or indirectly through *associations* with material artefacts (Latour 2005).

12.2 Donna: Integrated Curriculum Making in ITE as Uncharted Territory

This story is about my experience of facilitating an innovative assessment task in initial teacher education (see King 2014) from its birth to its death 4 years later. The task spanned traditionally siloed discipline areas and required our preservice teachers to undertake an inquiry in EEfS. Throughout this time I was the program coordinator for the Graduate Diploma in Primary Education course. I was part of a team of original designers for what became known as the integrated assessment task and oversaw the implementation of the task during the 4 years. Our hope was that the preservice teachers could transfer their own experiences of inquiry in EEfS to a primary classroom as well as change their attitudes and practices to reflect the development of a moral compass that considers living for sustainability. In reflecting on the process of curriculum making, I draw on email correspondences with students and academic staff, and my own diary entries and highlight the difficulties I encountered preparing preservice teachers to learn about a local sustainability issue through an integrated approach.

12.2.1 The Integrated Assessment Task

I was in a leadership position as the program coordinator for the Graduate Diploma in Primary Education course and a member of a group of lecturers who came together to address a problem that we had identified in the course. We felt that the assessment load was too heavy for our preservice teachers and that existing assessment tasks were uninspiring and repetitive. As the lecturers of three subjects within the course, we created a culminating task satisfying the learning outcomes of all of our subjects. This task challenged students to think about local sustainability issues and how they could be viewed through the various disciplinary lenses (i.e., Science/Technology, Arts/LOTE, SOSE). We designed the task to afford students opportunities to be reflective of sustainable practices in their local communities possibly leading to transformative practices in their own lives. We spent many weeks designing and refining the task that required students to devise a key question to frame their investigation, research the issue from the perspective of the five disciplinary areas, collect data from primary and/or secondary sources, analyse the data, propose an innovative recommendation/solution to the issue and produce a pedagogical resource. Finally, through a "Celebration of Learning" day, our preservice teachers presented the outcomes of their inquiry to their peers. Through the innovative use of digital technologies they described their original solutions to local sustainability issues. At the celebration day at the end of this first year, we were so proud of the preservice teachers when they demonstrated their deep knowledge with the issue and discipline areas, often describing the transformative practices that occurred in their own lives.

I led the implementation of the task for 4 years and while the outcome was often rewarding, the journey to reach this final celebration day was tumultuous and at times, emotionally exhausting. One challenge for me was the changing of the teaching team each year since the university's complex timetabling system could not prioritise stability of staff across the subjects.

12.2.1.1 Curriculum Making

My role required that I constantly negotiate with the teaching staff to ensure the explanations to preservice teachers about the task were consistent. Initially, I delivered a well-planned introduction so that preservice teachers were informed of the many benefits of the task. This included the value of doing collaborative inquiry as adults, the importance of learning to work in groups, the advantage of a decreased assessment load and the need to critically view science related to sustainability issues. After my introduction, I fielded questions from the whole group and answered questions via face-to-face and online modes. Despite many explanations, preservice teachers needed constant reinforcement about task requirements. My field notes show the confusion of preservice teachers' despite the many explanations:

> Some are trying to find the easy option…so I reinforced that they had to carry out the inquiry themselves and not just talk about it e.g., ring the local council, conduct surveys, research, do a science experiment and collect data and report on it etc… This seemed to be a surprise to some…who were not aware of the inquiry process. Some groups are off and running and others are still trying to decide on an issue. Most that have approached me have an OK idea but some are trying to make it too large e.g., renewable energy in Australia and I suggested they narrow it down to a local level (e.g., wind power is difficult to utilise in Brisbane). (Field Notes, August, 2014)

While I was committed to making the implementation a success, and convinced that the outcomes were worth the investment, I often felt fatigued from the emotional and administrative load required each year to set up the task and ensure that preservice teachers and the lecturers understood the expectations. When I received repetitive questions from preservice teachers throughout the semester, often with regard to how they approached the task I was left wondering "was it worth it?"

My other challenge was convincing the lecturing team of the value of the task, especially if they were new members. There was some resistance because they were concerned about the pressure of a 1-year course with inadequate time to address the discipline content knowledge for each curriculum area adequately. One example of this was in the SOSE unit where the teaching team member (Lecturer One) wrote this reflection:

> I felt that the preservice teachers completing the inquiry task were effectively skimming over a lot of the key content and skills they would need to be able to teach in-depth when they were practicing teachers. I was really concerned in my area of SOSE that that they would have NO idea that civics and citizenship, plus history were part of the curriculum – essentially, the inquiry task forced them to focus on sustainability at the expense of the other learning areas that comprised SOSE. I worried that after modelling integrated learning and assessment in this way at uni that the preservice teachers would implement units that did not teach key content and skills when they were employed as teachers. (Lecturer One's reflection)

Fortunately, after the lecturing team had experienced one iteration of the task where they had assessed students' oral work through the "Celebration of Learning" day as well as the written component, their appreciation of the learning that occurred for preservice teachers changed. One example of this occurs below in Lecturer One's reflection:

> As a teacher educator I did not fully appreciate that if integration is to be done well that we have to "let go" of key content in order to teach the concept. In the end, the students developed deep understandings of sustainability and how to plan and deliver integrated learning experiences, which I have to say was very worthwhile and prepares them well for the reality of primary school teaching. (Lecturer One's Reflection)

Many of the preservice teachers' perspectives changed also after they had completed the task and they were more appreciative of the learning process. One representative preservice teacher comment is below:

> Throughout the whole assignment I didn't think I would learn so much from group members and self-research. (PST Four)

12.2.2 *Viability of the Task and the Institutional Context*

While the institution awarded the teaching team a "Supporting/Enhancing Learning" award in its third year of implementation, the institutional arrangements continued to hinder the implementation of the task. Changing teaching teams meant new teams needed to be initiated into the philosophy behind the inquiry task year after year. Unsuitable class timetabling of the three subjects prevented the availability of all three lecturers concurrently, inhibiting opportunities for co-teaching classes and more united approaches to integration. I was not provided with any extra time in my workload to administer and coordinate the task and I am sad to report that this led to my exhaustion and diminishing motivation. After the fourth year of its implementation, the integrated assessment task met its death. The Graduate Diploma in Primary Education was superseded by the new Master of Teaching Primary Course. Unfortunately, I was too disheartened to be the champion for the task's inclusion into the new course after the struggles I had endured and the task was discontinued.

12.3 Jenny: Curriculum Making in ITE as Contested Territory

I am lucky to work with a team of critical educators known as the Science Transformative Learning Group. This is the story of the group's formation and about curriculum development in initial teacher education at the Australian Catholic University in Melbourne. However, this is not a story of curriculum development the 'right' way (Apple 2006). Rather, it is a story of disruption to a product-oriented curriculum review process at my institution. This curriculum review threatened to render critical approaches in my science curriculum-making work unviable.

Our course in science education is for Bachelor of Education preservice primary teachers. This undergraduate course, consisting of three semester-long science subjects, exposes our Bachelor of Education students over the 4 years of their degree to critical approaches, including Environmental Education for Sustainability (Martin and Carter 2018), Bioethics (Castano Rodriguez In Press), Indigenous Perspectives and SocioScientific Issues. I first started working in this course in 2002 as a sessional staff member and became full time faculty in 2012. Prior to my involvement, Lyn Carter and Caroline Smith had already established a successful course in critical approaches to science education and I was inducted into the critical paradigm through their mentoring. In 2011, Carolina Castano Rodriguez joined our science education team and developed the third subject in science education in this course to include partnerships with primary schools. In this subject, our preservice teachers now develop critical science curriculum units and teach these to small groups of students in the partnership schools.

I have been the lecturer in charge of the first-year subject of this course for the past 4 years. Beginning with the course I inherited from Lyn and Caroline, which used EEfS as a context for learning science, I have been involved in curriculum-making over these years as a largely autonomous process. In my first year of lecturing in the first year subject, I introduced an assessment task called the Eco Challenge, which Lyn and I have written about elsewhere (Carter and Martin 2018). In my third year of lecturing in the first year subject, I introduced the Ethics of Care, the background to which I will briefly explain although we have written about it elsewhere (see Carter et al. 2018).

Last year, Carolina and Lyn and I embarked on a pilot research program called the Ethics of Care project with a team of practicing teachers to explore the potential of Nel Noddings' ethics of care philosophy in primary science education for improving the science learning and engagement of children from a low socioeconomic part of Melbourne. Following the success of the Ethics of Care project (Castano Rodriguez and Martin 2015), I decided to introduce the Ethics of Care in the first year of our course.

12.3.1 Curriculum Making

First-year students were required to keep a personal reflective journal over the 12 weeks of the course, reflect weekly upon the EEfS course content, and to finally submit a report that used extracts from their journals to summarise their learning and relate it to eight 'Centres of Care' (Care for self, close and distant others, built and natural environments, other species, cultures and ideas, Carter et al. 2018). Our institution provided ethics clearance for Lyn, Carolina and me to collect the preservice teachers' journal entries as data for our scholarship related to the development of preservice teachers' pro-environmental engagement. I include an excerpt from one of the preservice teacher's journals as an example of how the preservice teachers' reflective writing is in turn providing me with data upon which to further my curriculum making in the course. I have chosen this excerpt because rather than it being a clear-cut example of success in the course, it provides information about the variety of beliefs our preservice teachers do hold. Better understanding of our preservice teachers is surely an important aspect of curriculum design for our critical agenda.

The journal excerpt, reproduced verbatim below, was published online by the preservice teacher using Tumblr.com. In the excerpt, she reflects upon the BBC documentary 'The Origin of Us', shown during my lecture on evolution, climate change and the extinction of species in week 10 of the course. In the excerpt, the preservice teacher attends to themes of anthropocentrism versus ecocentrism, scientific versus non-scientific views of species differentiation, and open-mindedness versus prejudiceness informing interactions with others.

Week 10: The Story of Us

In the documentary, the connection that we humans have with other species was discussed, as the presenter was talking about how we're not as "special" and different as we may like to believe ourselves to be because we're actually not *that* different to animals.

I found it really interesting when she was speaking about how we evolved from monkeys millions of year ago, and how our characteristics are very similar to them, such as the way we go about searching for food. Although confusing at first, I was really intrigued hearing about her reasoning regarding why she thinks we evolved from monkeys and how they are our ancestors.

She explained about how our ancestors, the "Homo Erectus" were hunters and that's how we evolved, and the diet's of our ancestors shaped how we humans are today.

All in all, I found the documentary to be quite compelling, and although it doesn't particularly correspond with my own views and what I've been grown up to believe, it was still very engaging and I'm still very open-minded to other people's views.

(BBC: The Origin of Us 2011)

What Centres of Care does this demonstrate?

The topic of how we humans originated is open to so much different interpretations, and I think this documentary showcased a "Care for Other Ideas, Beliefs and Cultures".

It reminded me to keep open-minded and not be prejudiced towards beliefs that might not align with my own.

(Reflective Journal accessed 20/01/2017, http://science-inourworld.tumblr.com)

The success that I see in this example is that the preservice teacher has had the opportunity to (re)position herself as open minded to scientific viewpoints, as intrigued by the scientific theory of evolution, and as a moral citizen and beginning professional open to the ideas and beliefs of others. Her repositioning is in relation to public readers of her blog, but also importantly, her lecturers and tutors, who are in effect gatekeepers to the teaching profession for her. In this way, she positions herself not only as a citizen sharing moral obligations with and towards others in society, but also as a beginning professional taking a stance on practices valued within a teaching profession (see Edwards 2017). Most of our undergraduate students come to the primary teaching course having not studied science beyond the compulsory years of schooling. Some hold beliefs about creationism that are incompatible with scientific views. However, when they become teachers, they will be obligated to teach science according to the Australian curriculum. Not only has this preservice teacher developed a better understanding of the scientific concept of evolution but she is also developing a capacity and willingness to engage in conversations about the views of others, including scientists, and hopefully also the views of her future students.

12.3.2 Institutional Processes

Curriculum making also occurs at an institutional level. In our institution, each course is subject to a 5-yearly cyclical review (ACU 2015). In 2014, the review process began at another campus of our institution. We received a rewritten course that would have made our critical approaches completely unviable, requiring us to teach science as discrete discipline areas, focusing entirely on scientific concepts that would be tested in a traditional way using recipe practical work and concept regurgitation. Not only would we lose our intellectual freedom in curriculum making but also our research program would be on the scrap heap. Over the period of 2 weeks in the month of July 2014, we enlisted ourselves into the course review process, firstly by objecting in a collectively-written email to the National Director of Primary Courses, and secondly, by facilitating working parties to rewrite the course documents. The National Director enabled generic documents to be written that would maintain our intellectual freedom moving into the future.

The official formation of our group, the Science Transformative Learning Group occurred soon after the course review process. Whilst we had been working as a team for over a decade, the naming of our group, like the claiming of our territory seemed important and necessary to us. We created a glossy flyer, claiming our group identity and describing our goals and achievements, which we sent to faculty members to create better within-faculty awareness of our critical approaches and since used in correspondence with outside and partner organisations for research proposals, reports and other correspondence within and beyond our institution (e.g. Castano Rodriguez and Martin 2015).

12.4 Maintaining Our Critical Work

Donna reports on changes in ways of talking about preservice teacher competencies as a result of preservice teachers' and academic colleagues' participation in an integrated assessment (inquiry) task that she had designed so that preservice teachers could experience transformative science education positioned as adult learners. Prior to the experience, preservice teacher competence was associated with lesson plans and knowing and incorporating key content into lessons. Through the experience, the learner-centred approach modelled by the integrated assessment task engendered talk about the necessity of "letting go of key content" and repositioning the preservice teacher/learner as agentic.

The idea of uncharted territory suggests the need for navigation. Indeed, we see the idea of navigating a new space playing out in Donna's story. The responses to the task from both the preservice teachers and the lecturers were a surprise to Donna in the sense that she found them diary-worthy and recorded her reactions to the questions and concerns they raised. Donna experienced the initial reactions of the preservice teachers and some faculty members as resistance to the task. It became clear to Donna that others' expectations of the journey through this territory were different to hers. To the preservice teachers and other lecturers this territory was new. The curriculum making work in this story included the mapping of the territory for others, an imaginary including the purpose of the task and the affordances within it for learning and change, and jointly navigating through this new territory. In each year of the iteration of the task, the preservice teachers and lecturers needed to step on board for an exploration of new ways of being and enacting in the world of initial teacher education.

Donna's story clarifies the work needed to facilitate change in this way. Maintaining her critical approach in this example required better recognition and provision from her institution for her leadership as well as better communication from Donna about staffing needs. Had she been able to choose suitable faculty members, whose visions for ITE aligned better with her own and with the goals of the task at the outset, and been able to maintain a consistent team across the iterations of the course, then the viability of the course may have been significantly enhanced. It is short sighted in an institution to fail to value and enable leadership in innovative curriculum making.

Jenny and her colleagues' values in curriculum making were initially passed over in a product-orientated course review process. She reflects upon enlisting herself in the course review process to protect her intellectual autonomy. An outcome of the process, an emotional experience to say the least, was the claiming of a group identity: the Science Transformative Learning Group. As a member of the science Transformative Learning Group, her work in curriculum making in ITE using critical approaches continues.

The contested territory metaphor suggests conflict. However, this is not what was experienced. Rather, a lack of concern or awareness of the existence of different approaches in curriculum making for initial teacher education in science and an undervaluing of the importance of academic freedom was accompanied by or brought into being through time pressure to get the job of the course review done. Ultimately, once Jenny and her critically-minded colleagues held their ground collectively, their practices were acknowledged, valued and enabled.

Together, our stories show some of the passion and labor involved in maintaining what we value in our work in initial teacher education and clarify for us the relational aspects of such work. Curriculum making is shown as a process of negotiation with others in its enactment with preservice teachers and other faculty members. However, our stories also highlight the institutional level of curriculum making that occurs as the performation of institutional practices, whose transformation is often seen as beyond the capacity of the individual. Here, our growing awareness of the importance of building relationships and working with others to create spaces within which we can engage in critical reflection and action to make the resolution of contradictions in our institutional arrangements possible is shown to be crucial for maintaining our valued work.

References

Apple, M. (2006). *Educating the 'Right' way: Markets, standards, god, and inequality*. Hoboken: Routledge.

Australian Catholic University. (2015). *Quality assurance handbook*. https://handbook.acu.edu.au/handbooks/handbook_2015/general_information/assessment_policy/11._quality_assurance

Carter, L., Castano-Rodriguez, C., & Martin, J. (2018). Embedding ethics of care into primary science pedagogy: Reflections on our criticality. In J. Bazzul & C. Siry (Eds.), *Critical voices in science education research: Narratives of academic journeys*. Dordrecht: Springer.

Carter, L., & Martin, J. (2018). "I had to take action straight away". Pre-service teachers' accounts of pro-environmental engagement. In J. L. Bencze (Ed.), *Science & technology education promoting wellbeing for individuals, societies & environments*. Dordrecht: Springer.

Castano Rodriguez, C. (In press). Challenging speciesism: Youth repositioning of identities as ethical adults. In G. Reis, M. P. Mueller, R. A. Luther, L. Siveres, & R. Oliveira (Eds.), *Ociocultural perspectives on youth ethicalonsumerism*. Dordrecht: Springer.

Castano Rodriguez, C., Martin, J. (2015). *Ethics of care in science*. Report to Catholic Education, Melbourne.

Edwards, A. (2017). The dialectics of person and practice: How cultural-historical accounts of agency can inform teacher education. In J. Clandinen, & J. Husu (Eds.), *International handbook of research on teacher education*. SAGE.

King, D. (2014). Preparing pre-service teachers to teach primary science: An integrated approach using the theme of sustainability. *Teaching Science, 60*(4), 34–37.

Latour, B. (2005). *Reassembling the social: An introduction to actor-network-theory*. Oxford: Oxford University Press.

Martin, J., & Carter, L. (2015). Preservice teacher agency concerning education for sustainability (EfS): A discursive psychological approach. *Journal of Research in Science Teaching, 52*(4), 560–573. https://doi.org/10.1007/s11422-012-9412-4.

Martin, J., & Carter, L. (2018). "Finding ways to fill the void": Preservice teachers struggling with their own consumption. In G. Reis, M. P. Mueller, R. A. Luther, L. Siveres, & R. Oliveira (Eds.), *Sociocultural perspectives on youth ethical consumerism* (pp. 19–32). Dordrecht: Springer. https://doi.org/10.1007/978-3-319-65608-3_2.

Mitchell, C., Weber, S., & Yoshida, R. (2008). Where are the youth?: Reframing teacher education within the context of youth participation. In A. J. Sumsion (Ed.), *Critical readings in teacher education: Provoking absences* (pp. 139–154). Rotterdam: Sense Publishers.
Stetsenko, A. (2014). Transformative activist stance for education: The challenge of inventing the future in moving beyond the status quo. In T. Corcoran (Ed.), *Psychology in action* (pp. 181–198). Sense: Rotterdam.
Tilbury, D. (1995). Environmental education for sustainability: Defining the new focus of environmental education in the 1990s. *Environmental Education Research, 1*(2), 195–212.

Jenny Martin (previously Jenny Arnold) is a lecturer in science education at the Australian Catholic University in Melbourne. Her research is ultimately concerned with learning science for just and sustainable societies. She won the Australian Association for Research in Education 2017 Early Career Researcher Award for her paper entitled, 'Researching Preservice Teachers' Environmental Agency in Times of Post Truth and Denial'. Her research on student, teacher and preservice teacher agency using discursive psychology has been published in the *Journal of Research in Science Teaching*, *Learning Culture and Social Interaction*, the *International Journal of Science Education* and *Research in Science Education.* She has made invited contributions to forums for *Cultural Studies of Science Education* and co-guest edited a special issue on student agency in learning interactions for *Learning Culture and Social Interaction.* She is a convenor of the Sociocultural and Activity Theory Special Interest Group for the Australian Association for Research in Education and in this role facilitates mentorship and networking for higher degree students and early career researchers.

Donna King is an Associate Professor in science education at the Queensland University of Technology. Her research in science education spans three interconnecting fields: the emotional engagement of students in science in the middle years, using engineering contexts for teaching science and context-based science education. An outcome of this work is the development and implementation of innovative context-based units where teachers have adopted new pedagogical approaches for teaching science. Recently, she completed a project with a team of researchers from the Queensland University of Technology and the Department of Education and Training to establish a STEM (Science, Technology, Engineering and Mathematics) Hub for schools where teachers worked with industry partners to connect STEM in the classroom with real-world STEM. She is the co-chief editor of the international journal *Research in Science Education.*

Chapter 13
Confronting Self: Stories of Incipiency, Disequilibrium, and Becoming Critical in Science Education

Darren Hoeg, Larry Bencze, Sarah El Halwany, Erin Sperling, and Majd Zouda

In this chapter, we present stories related to critical scholarship stemming from our research in science and science education. We contribute this chapter together, first and foremost because each of us has strong 'critical' beliefs and perspectives that we want to 'do their work' in the social settings with which we engage, and, more generally, in society at large. What critical means may be different for each of us, and was the topic of vigorous discussion, without consensus, during planning meetings for this chapter. At their core, however, our critical views center on practices of *citizenship*. We see citizenship as, ideally, a participative, socially-constructed and dynamic subjectivity, rather than conferred status, in which individuals make decisions on, and challenge, the structures of society. We are concerned about what Henry Giroux (2008) terms a 'hollowing out' of civic life, and subsequent colonization of citizenship by economic, market-based rationalities and practices. Hollowing out, claims Giroux, results from the lack of citizen participation in civic/social activity, such as voting, social activism, and community-oriented decision making. This is exacerbated by the prioritization of individual rights and obligations, often directed toward self-investment and advancement at the expense of the common good. We have many questions about how school science may contribute to the hollowing out of social life in ways that are poorly aligned with democracy and community values, and we are committed to research aligned with this theme. While writing this chapter, we frequently discussed our opposition to certain common, dominant beliefs and values in science education communities, which we termed the "mainstream". Although we are hesitant to construct boundaries, it seems difficult to argue that there is not a large, perhaps majority, community of mainstream

D. Hoeg (✉)
University of Alberta, Edmonton, Canada
e-mail: dhoeg@ualberta.ca

L. Bencze · S. El Halwany · E. Sperling · M. Zouda
University of Toronto, OISE, Toronto, Canada

J. Bazzul, C. Siry (eds.), *Critical Voices in Science Education Research*,
Cultural Studies of Science Education 17,
https://doi.org/10.1007/978-3-319-99990-6_13

science educators who prioritize what are often termed "high status" knowledge and practices in science education (Apple 2004). This high-status learning includes the acquisition of the established knowledge and facts of the discipline, and the rehearsal and performance of standard laboratory practices of science, which are advantageous to already privileged communities, but likely inaccessible to the majority of students (Hoeg and Bencze 2017). These requirements appear to be maintained by *gatekeepers* (Mueller 2011) in science education who try to conserve what counts in terms of impact in the classroom and what counts, or not, for the purposes of doing good work in science education. Gatekeepers, often associated with high status journals in science education, wield power to grant or refuse access to the field by, among other ways, rejecting research submissions. Rejection occurs particularly when scholarship submitted does not conform to high status epistemological norms that are prioritized by gatekeepers, such as the positivism central to much quantitative research. We see our values and beliefs as different than those of gatekeepers, positioning us outside the mainstream. The stories that follow describe some of the tensions each author incurred attempting to reconcile perhaps more critical perspectives with mainstream expectations.

We also write this together as individuals in various relationship to each other; we are colleagues, working in the same institution; we are science education scholars; we are friends; we are supervisor and supervisees. In these roles, we have supported each other in conducting critical scholarship when facing rejection from mainstream science education communities, which can take an emotional toll (Butler 2004). Each of our stories might be described as an experience of *becoming* a critical scholar. Philosophies of becoming have existed since ancient Greece, and generally refer to processes of change and "moving towards", presumably some underlying or "true" reality or state, which may be difficult to perceive through human sensing of the material world. In more contemporary philosophical approaches, the notion of becoming includes the creation of ways of coming to knowing previously unknowable reality, as new perceptions of reality occur with/in the *self* (Conolly 2013). Although the concept of self can be explored from philosophical, psychological and sociological perspectives, its description as the organized, consistent set of perceptions and beliefs about oneself (Rogers 1961) aligns with our usage of the term. The self can be thought to be formed through its relation to "objects"; the constructions of the human mind that represent reality, as we come to know it, which are imbued with recognizable and definable characteristics perceivable by the human subject (Sewell 1992). The stories below represent particularly relevant and transformational personal experiences in which we come to better understand how self shapes, and is shaped by, the research that we do. Our stories revolve around experiences of self-transformation – that is, new understandings of self, due to various powerful experiences while engaged in academic scholarship or research. Although each story is unique, they hinge on events that evoke deep personal tension, reflections on self-beliefs, and adaptations of scholarship. Our stories, hopefully, provide points of reflection for others, in similar circumstances, to advance critical voices to social settings that may most benefit from them.

13.1 Sarah

I embarked on my PhD anticipating that I would graduate feeling 'transformed'. I expected this transformation to occur unproblematically; that is, a smooth process in which I simply came to see the world from a different lens. Yet, I am discovering that my transformation has occurred not merely by directing my academic gaze outwards, to society, but also through reflection on my inner personal assumptions, values and beliefs. Many of these beliefs are derived from my background; I am a recent immigrant who grew up in a relatively conservative and religious family. My conservative beliefs have produced tensions with some of the ideas and perspectives that are part of the critical community each author of this chapter is engaged in. Voices that advocate for social and environmental justice have at times sounded paternalistic, "telling" me where to focus my attention, what to think, believe and feel. Was I resisting being "molded"again? Or is it that I am unwilling to leave the comfort of my own mold shaped by gender, culture and social traditions? Regardless of underlying reasons, I can't help but feel restricted in what I can wholeheartedly advocate for and thus find it even harder to locate myself in science education scholarship. Thus, participating in a critical culture has been an ambivalent experience, empowering at times, and at others incurring a feeling of being displaced, as if I am always "in-between" spaces (Aoki 1993). This unrest is often compounded by my inconsistent relationships to different beliefs and values of self. Another way to think of this is as a *disequilibrium*. In this chapter we consider disequilibrium to be a kind of ontological and epistemological uncertainty resulting from awareness of new values and beliefs, stemming from new perceptions of the world, that may be in active opposition to prior beliefs (Connolly 2013).

My disequilibrium has been further reasserted in various academic encounters. At a NARST conference, I found myself amongst science education researchers talking enthusiastically about the topic of my presentation related to using socio-scientific issues (SSIs) to teach about the complex science and technology networks involved in our consumer-based economy. SSIs can be very broadly defined as controversial issues related to applications of science and technology, such as climate change, and development of genetically modified organisms. The general reaction to this work was, "but how is this science education?" Everyone there seemed to agree that political and economic dimensions of socio-scientific issues should be discussed in social studies classes but not the science classroom. Something in me wanted to concur. This perhaps stemmed from changing, yet still influential beliefs that science is a sanitized and pure subject. Rather than engaging with these familiar and comfortable beliefs while discussing my topic with the science educators present, I did not object. As a result, I felt a form of 'guilt' from my acquiescence to the more conservative beliefs about science education that came with my silence.

In a yet another salient recollection, I attended a session- during the European Science Education Research Association conference- in which some members of the audience expressed their concerns about the presenter's choice to use gender as a fixed independent variable. A voice in me surged and sought to justify the presenter's choice rather than to problematize it, as would be expected from critical scholars. At the end of the presentation, I found myself among the "critical" commentators that raised the question of the appropriateness of categorizing gender. While I could have presented some well formulated academic arguments to join my voice with them, my uncritical voice took over, justifying once again, this time outloud, the presenter's findings in ways that re-inscribe normative gender behaviors. Needless to say, those critical commentators were instantly dismissive of my remarks. For them, I was perhaps seen as uncritical, or unfit of criticality. For me, it was the crux of my being in-the-middle, wondering whether my criticality will ever be encompassing, ever achieved.

13.2 Majd

My experience as a Ph.D. candidate in OISE has resulted in significant changes in my views about many things, including the notion of criticality itself. Development of my critical perspectives has been greatly advanced through work towards my doctoral thesis, in which I focus on how integrating Science, Technology, Engineering and Mathematics (STEM) education is conceptualized and practiced in some Canadian high-schools. There have been some recognizable efforts to construct STEM education to be inclusive to diverse students, and directed toward community and democratic needs. However, dominant views and practices appear to be confined by economic goals, resulting in emphasis on instrumental skills and knowledge for training purposes (Gough 2015), lack of support for ethical, active citizenship (Zeidler 2016), and employing STEM education as a means to advance transnational for-profit agendas (Hoeg & Bencze 2017). While acknowledging the valuable potential of STEM education, I embraced these critiques, navigating away from the mainstream STEM.

The initial setting for my research was intended to be a school board in Ontario, with a new STEM program. At first, they welcomed research collaboration. However, after I explained the critical lenses through which I approach STEM to the program coordinator, their enthusiasm for research collaboration disappeared, and they instead decided that they are not ready to be studied. The timing and nature of their response suggested anxiety toward the critical nature of the study. This result was distressing for me, and caused a *retreat*. Retreat was an occurrence experienced by many of my coauthors, after a particularly difficult experience that forced a reflection on our self-beliefs. Retreat came to be how we described the period of time after self-transformative experiences, in which we re-evaluated our beliefs and positions, often resulting in changes to our existing beliefs, and/or new approaches to our scholarship. Retreat resulted in my most 'critical' moments, as I re-evaluated the lenses through which I approach STEM education. However, through this contemplative search, I regained confidence in my beliefs – that dominant constitutions of STEM education tend to be narrow in focus, inaccessible, and therefore unlikely to provide claimed benefits to many students. These beliefs, I have realized, are foundational to my commitment to forms of STEM education (and research) that are useful to the majority of students, who need skills to evaluate, be critical of, and act, on STEM knowledge and practices, for the betterment of communities and the common good.

At the same time, I could not but acknowledge the right of others, particularly research participants, to differently perceive STEM (and the world). When I could clearly define my position, I had to negotiate the dilemma of recruiting research participants while remaining sincere to my own beliefs. A colleague of mine advised me to be careful when framing the purpose of my research and my underpinning beliefs to participants. However, although I acknowledge the need for nuanced negotiation between the researcher and researched, I feel bound by ethical obligations for transparency with my research participants. I also feel anxious about the consequences of producing findings that do not align with my participants' views or are critical of their practices: other than the risk of withdrawing themselves and their data from the research, the need I feel for trust in the researcher-participant relationship drives me to think on how to sincerely present my research findings without 'offending' my participants.

If we acknowledge that the research process is a series of negotiations and interpretations of realities (Denzin and Lincoln 2000), then every stage represents a negotiation between the reality we embrace, and the perceived realities of research participants. Although all researchers may face such dilemmas, having perspectives that sharply depart from the mainstream, which is often embraced by participants, may increase and intensify this possibility. At each stage, critical perspectives may be perceived as a potential threat by participants', to their perspectives, practices and/or their ways of being, increasingly isolating the researched from the researcher. Therefore, not only recruiting participants would be challenging, but also expecting them to truly expose themselves or approve research findings could be compromised.

As a possible solution, I decided to have a section in my research representation dedicated for participants' interpretations and counter arguments. Meanwhile, there are questions I am still reflecting on as I progress as a critical scholar: How can we, as 'critical' researchers, maintain our authentic voices, establish our niches, and proceed with research while negotiating the self, others and various possibilities? Does this represent a reversal of criticality, or criticality in one of its most sincere forms?

13.3 Erin

For me, being critical means, in part, enabling research participants to engage in place-based, research-informed activism, typical of Participatory Action Research (PAR). Rather than a more traditional observation, PAR involves facilitation of participants to inquire about, create data from and propose and enact actions toward local, community-based change. My doctoral research looks at non-formal education settings, as these are thought to have fewer curricular requirements than schools, that tend to constrain student activism (Bencze, Sperling and Carter 2012). My initial intention was to engage youth in PAR on socio-scientific or environmental issues in their community.

I was really excited about the prospect of working with youth in a PAR project at a food justice organization in an urban centre. The organization was initially supportive of this collaboration, and began discussing possible student-led actions with me, but due to limitations of time, human and physical resources, and the commitment of youth participants to the other aspects of their after-school program, it became evident that PAR was not possible. The program facilitator and other community-engagement staff explained to me that activism, while embedded in the mission of their programming, was not possible for the youth participants because of their desire for a slow and scaffolded process for community activism. For me, there was no option but to honour their request. From this experience I became aware of tensions inherent to criticality as a social justice project, potentially causing these projects to backfire on the ground, if the intended participants do not feel equipped for taking social action, or their desires are not being honoured. How can I, as researcher, help my participants further their work and their sense of agency, if they do not feel ready? I began to recognize that my initial idea for PAR research would have required a demonstration of power and privilege, and forms of agency, that the youth at the urban center likely did not have, given the short commitment time in the afterschool program, as well as realities of socio-economic and linguistic barriers.

Still seeing great value in the work of the site and the research I could conduct there, I 'retreated', to reconsider how to advance a research project with/in this community. My research program, and my own beliefs and values, had to shift to accommodate desires and perceived potential of the research participants. I decided to conduct an ethnography, which allows for critical perspectives on the part of the researcher, but does not necessarily engage participants in critical knowledge-production activity as explicitly as PAR does. This shifted methodology, however, presents ongoing challenges of being critical of my own position as a privileged body/researcher in the community. How can we "do" social justice research from the space of dominance or from places of privilege? Being critical in this renewed research approach of ethnography involves making visible the power structures oppressing participants, so that I may identify relevant justice-oriented social activity, slowly shifting and dismantling oppressions from within. For example, I am able to observe agency in their work toward bettering themselves and their communities, in activities such as intergenerational gardening and healthy food consumption.

This experience also demonstrates the border-crossing that had to take place for me to maintain my position as a co-facilitator and a participant-observer. I had to leave much of my academically-formed assumptions behind to engage in the research as it was reimagined, such as my assumption that research-informed activism was the best fit for my research site. I hope that at a minimum my awareness of this challenge will help me to approach deeper understanding as I report on my research. My resilience as a researcher is ongoing, as I continue to develop entry points and moments of deconstruction of self as researcher, within the anticipated outcomes of ecojustice education. I continue to try to provoke new perspectives in science education that allow participants to feel grounded in their own knowledge building, by starting where they are ready and joining them on their journey.

13.4 Darren

Over the summer of 1998, I conducted scientific research for Agriculture Canada, to determine if honey bees were pollinating indigenous low bush blueberry. My task was to obtain and identify the pollen collected by bees from nearby fields in the Annapolis Valley of Nova Scotia. Arriving at the blueberry fields each day, I was encompassed by the resonant sound of thirty to forty thousand busy and seemingly content animals, the din reminiscent of a vibrant note from a large pipe organ in a church. Wearing a protective beekeeping suit, and thick leather gloves, I squatted near a hive entrance, inundating the hive with smoke, which acts as a mild sedative, so I could collect the bees. I gently reached for a bee, grimly discovering that the cumbersome leather gloves made it difficult to avoid maiming or destroying the insect. I grasped a second bee more gently, but still unintentionally killed it. The sound of the colony increased in pitch and intensity, becoming alarming, the low pipe organ reaching a higher octave. I was shaken as I felt sharp impacts through my suit; the irritated bees had organized an aerial assault to attack me. Over the next 10 min, hundreds of bees gave up their lives to eradicate the threat they detected in their colony, finding openings in the suit, or driving their posteriors with enough force to sting through the thick garments I wore.

I felt as an alien intruder in an uninviting world. A reversal of roles, from predator, to prey, and growing awareness of my connection to the bees as living parts of nature, allowed me to sense previously unrecognizable aspects of the life of the bees. A form of intelligence, and group consciousness, phenomenon difficult to observe through the reductive scientific sampling procedures I was using, became knowable. The experience resulted in fractures within my understanding of self, an understanding largely based on a notion of being separated from nature. Separation from nature allowed me to be an 'objective' observer of the 'other' (nature). This deeply personal experience illuminated an *incipiency* in my understanding of nature. Incipiency can be thought of as a recognition of impending transformation of understanding, a sense of potentiality in epistemological boundaries, often resulting from transformative personal experiences (Connolly 2013). My experiences with the bees revealed a possibility of knowing the other (nature) in ways I couldn't perceive from objective and positivistic epistemological and ontological vectors, measuring and quantifying nature, that are important in science. As my *scientific-self* became dislocated from this experience, I felt uncertain of the 'reality' of the bees, and unsure how to understand their reality. The value I had for the bees changed, from one of scientific utility, to reverence for these powerful living organisms with their own agency and purpose. But what exactly was the agency, what was the unknowable purpose of the bees? I spent years trying to understand this awakening of other aspects of self, which progressively became attuned to how Eurocentric scientific ways of knowing may shape how it is possible to know nature, and critical of how other ways of knowing nature, both extant or as of yet uncreated, are marginalized.

I left the field quickly that day, chased for several miles in my car by the bees. The bees have attained symbolic importance for me, representing unknown ontology of the world, perhaps knowable through emerging epistemological practices, or extant marginalized practices, such as Indigenous ways of knowing. Learning outdoors, for example, through creative epistemological approaches, such as artwork, meditation, and spiritual practices, although typically not seen as appropriate in science education by gatekeepers (Mueller 2011), may allow students to experience incipiencies in understanding nature. These incipiencies might lead to expansions in the scope of science, in which nature is known not as 'other', but as part of the same fabric as, and through a singular ontology of which, humans are a part. Changed understandings of nature resulting from such experiences are necessary, I believe, to expand the scope of science/school science in advanced capitalist societies, so we might live sustainably with/in nature.

13.5 Larry

Being a science educator and researcher has, for me, largely felt like swimming against the current. Many of my most prized perspectives and practices seem antithetical to those promoted around me by colleagues, government and school district officials, textbook publishers, teachers, school administrators, students and others. Although it is difficult to pinpoint when such discontinuities began, prominent in my mind is frustration I felt early in my career towards opposition to my promotion of student-led primary research – including relatively uncommon correlational studies – leading to conclusions determined by students and possibly contradicting mainstream science education. After some initial puzzlement to explain resistance to such activities that I believe to be very agentic, it became apparent to me – particularly through reading books like "The Cancer Stage of Capitalism" (McMurtry 1999) – that such opposition may be due, at least in part, to influences of globalizing neoliberal networks. Neoliberalism is an ideology that, while a widely-accepted definition is lacking, appears to involve rallying of vast arrays of resources and contributors (e.g., transnational trade agreements, transnational organizations [e.g., Organisation for Economic Co-operation and Development], banks, think tanks [e.g., Atlas Network], financiers, universities, etc.) into a 'team' apparently promoting (e.g., via de-regulation, tax reductions, etc.) private sector interests (Springer, Birch and MacLeavy 2016). As discussed elsewhere (Bencze et al. 2018), these networks appear to me to be like *The Borg* (from the Star Trek™ programmes) – a menacing cyborg-like cooperative threatening to assimilate everything and everyone and, like a cancer, wreaking personal, social and environmental havoc along its path. Such oppression and damage deeply concerns me and motivates me to act for a better world. Accordingly, my prime educational goals have including efforts to encourage and enable citizens to develop and implement informed and negotiated socio-political actions to try to rectify relevant harms perceived by them. In part, this goal has been reinforced by Albert Einstein's (Calaprice 2000) advice:

> The aim (of education) must be the training of independently acting and thinking individuals who, however, can see in the service to the community their highest life achievement.

Although I have reported some successes in achieving these goals (e.g., Bencze 2017), ever-adaptable neoliberalism seems to have installed 'speed bumps', such as STEM, inhibiting progress in this regard. STEM appears to prioritize selection and training of a relatively small fractions of student populations that may become for-profit knowledge (and, more particularly, commodity) producers (or marketers, etc.) working as STEM professionals, while simultaneously indoctrinating most students to serve capitalists as knowledge *consumers* as, for example, enthusiastic, repeating and unquestioning purchasers of often non-essential commodities (e.g., Pierce 2013). Given such serious reservations about STEM education projects, it has been disheartening to me that most of my immediate science education colleagues have embraced this movement – particularly in pursuing institutional collaborations with engineers, focusing on development of innovations, apparently often without significant concerns about associated personal, social and environmental harms. While I largely attribute colleagues' orientation towards engineering-focused STEM education to hegemony of global for-profit systems, it seems that this may also be linked to some science educators' isolationist perspectives about fields of science and technology. It was startling to me, for instance, to hear colleagues' claim that scientists and engineers largely operate strictly in terms of logic and evidence, immune from political and/or economic pressures. Such a claim seems contradicted by case studies indicating numerous compromises to integrity of work of scientists and engineers contracted by government-sanctioned financiers (Mirowski 2011). Perhaps protection of images of integrity of STEM fields blinds attention to adverse outside influences.

While it has been relatively lonely working in a milieu apparently engulfed in neoliberalism-informed perspectives and practices, I am very grateful to work with supportive graduate students and, especially, to have a virtual community of like-minded scholars (many having chapters in this book) located in different parts of the world. It seems that perspectives and practices like ours are a 'mile wide and an inch deep.' Ironically, perhaps, we have found each other and maintained our community through capitalist infrastructure. In the context of our collaborative protests regarding an 'international' conference held in an exclusive gated resort, for instance, some of us formed a scholarly activist collective, using various Internet-based resources to sustain our project between conferences. Accordingly, it seems we have, likely in complex ways and for complex reasons, largely avoided assimilation into the neoliberal Borg, persisting in struggles for social and environmental justice.

13.6 Resolving Incipiencies, Gaining Equilibrium, and Continued Inquiry About Self

As our stories demonstrate, we have moved towards criticality along different, often complex paths. Yet, similar tensions were palpable components of each of our stories, and influential in our decisions as critical scholars. These tensions are deeply personal, and hint at transformations of self, evident in incipiencies and disequilibrium in our perceptions and beliefs about reality. Although we suggest no particular relationship or progression of these transformative events, an impending sense of changing perceptions of reality, or incipiency, was often an initial part of self-transformation. For example, Darren's experience with the bees revealed to him impending changes to his understanding of 'reality' (nature). Erin's story suggests an incipiency related to her understanding of the limitations of agency and power of low SES youth as she attempted to engage them in a PAR project at a local urban center. Many of the stories also suggest feelings of disequilibrium, an event related to incipiency, in that they both appear to be processes involved in transformations of self. For example, Sarah admitted to ongoing tensions in her existing beliefs about including socio-scientific issues in school science, a disequilibrium that manifested as a sense of in-between-ness. The frustration Larry felt early in his career towards the opposition of colleagues to many of his most prized perspectives and practices appears to be evidence of a disequilibrium that, upon reflection, caused him to seek communities in which he could achieve equilibrium in beliefs and perceptions of self and society. Majd and Erin express experiences of disequilibrium of self, related to research ethics; in each case, tensions in research settings invoked ethical and compassionate perceptions and beliefs, resulting in new awareness of self and participants, which allowed them to accommodate these new social realities.

Our stories appear particularly similar, in that, at initial stages of scholarship there was a degree of naïve expectation that our views should be unproblematically accepted. This might be explained by the very local-ness of the critical space we share at our institution, which isolates us to a degree from the scrutiny of others who may not share our critical perspectives. Perhaps propelled by confidence in our self-beliefs and perceptions, each of us enthusiastically entered into research engaging in practices representative of specific beliefs, such as Erin's initial beliefs related to equality of participants' ability to engage in activism, or Darren's somewhat positivistic epistemological beliefs that nature is knowable through objective description and classification, before interacting with honey bees. In each of our stories, initial rejection of our views by the subject of the investigation forced a temporary retreat, to understand our changing beliefs, transformations of self, and re-evaluate on how scholarship might then proceed. Retreat and reflection is clear in the questions Sarah asks about how to locate herself in research, considering her concurrent alignment to seemingly oppositional self-beliefs. Adjustments Majd made in her approach to recruiting participants, the shift in intent and methodology, from PAR to ethnography in Erin's study, and Darren's decision to leave science after his transformational experience with the bees, also occurred after a retreat. Retreat appeared to be a necessary period of time in which new ideas, beliefs and values about the world developed, and creative ways to utilize emerging beliefs and values to understand the world were imagined.

Far from being infrequently experienced and isolated events of our stories, early stages of planning this chapter also resulted in feelings of incipiency and disequilibrium, which emerged from reflecting on and questioning what critical scholarship is. The terms critical and scholar, for example, were seen by us to be somehow in conflict; critical suggested a resistance to certain dominant and/or privileged discourses of Academia, while advocating for oppressed 'voices'. Scholar implies to us a position granted to certain individuals, that is generally recognized by conformity to certain discourses that are dominant and privileged of a field of study. A perceived incommensurability between these selves resulted in more questions about how to be critical scholars than answers, such as: What should the outcome of critical scholarship be? What practices represent criticality, and just why are these critical? To what extent can one be critical and remain a scholar? Careful reflection on these questions challenged many of our fundamental beliefs about what it means to be critical, and required reflection of self, and retreat, for each of us to understand how to proceed in writing this chapter. These very questions are perhaps instances of the critical enterprise, not only posing a problem – what is this critique that we supposedly do or, indeed, aspire to do? – but enacts a certain mode of self-inquiry that we believe is central to the activity of critique itself. Far from answering these questions, instead, is recognition that this period of retreat, of being "critical about criticality", may be essential to understand how to negotiate self with other, learn from disequilibrium and incipiency, and enact critical scholarship.

References

Aoki, T. T. (1993). Legitimating lived curriculum: Towards a curricular landscape of multiplicity. *Journal of Curriculum and Supervision, 8*(3), 255–268.

Apple, M. W. (2004). Series editor's introduction. In W. M. Roth & A. Calabrese Barton (Eds.), *Rethinking scientific literacy*. New York: Routledge Falmer.

Bencze, J. L. (2017). Science & technology education promoting wellbeing for individuals, societies & environments. Dordrecht: Springer.

Bencze, J. L., Sperling, E., & Carter, L. (2012). Students as socioscientific activists: Impetus from their primary & secondary research. *Research in Science Education, 42*(1), 129–148. https://doi.org/10.1007/s11165-011-9260-3.

Bencze, L., Reiss, M., Sharma, A., & Weinstein, M. (2018). STEM education as 'Trojan horse': Deconstructed and reinvented for all. In L. Bryan, & K. Tobin (Eds.), Thirteen questions in science education (pp. 69–87). New York: Peter Lang.

Butler, J. (2004). What is critique? An essay on Foucault's virtue. In S. Salih (Ed.), *The Judith Butler reader* (pp. 302–322). Oxford: Blackwell.

Calaprice, A. (2000). *The expanded quotable Einstein*. Princeton: Princeton University Press.

Connolly, W. E. (2013). The 'new materialism' and the fragility of things. *Millennium: Journal of International Studies, 41*(3), 399–412. https://doi.org/10.1177/0305829813486849.

Denzin, N. K., & Lincoln, Y. S. (2000). The discipline and practice of qualitative research. In N. DenziN & Y. Lincoln (Eds.), *Handbook of qualitative research* (2nd ed., pp. 1–28). Thousand Oaks: Sage.

Giroux, H. (2008). *Against the terror of neoliberalism: Politics beyond the age of greed*. New York: Paradigm.

Gough, A. (2015). STEM policy and science education: Scientific curriculum and sociopolitical silences. *Cultural Studies of Science Education, 10*(2), 445–458. https://doi.org/10.1007/s11422-014-9590-3.

Hoeg, D. G., & Bencze, J. L. (2017). Values underpinning STEM education in the USA: An analysis of the Next Generation Science Standards. *Science Education, 101*(2), 278–301. https://doi.org/10.1002/sce.21260.

McMurtry, J. (1999). *The cancer stage of capitalism*. London: Pluto.

Mirowski, P. (2011). *Science-mart: Privatizing American science*. Cambridge, MA: Harvard University Press. https://doi.org/10.4159/harvard.9780674061132.

Mueller, M. (2011). Ecojustice in science education: Leaving the classroom. *Cultural Studies of Science Education, 6*(2), 351–360. https://doi.org/10.1007/s11422-011-9333-7.

Pierce, C. (2013). *Education in the age of biocapitalism: Optimizing educational life for a flat world*. New York: Palgrave Macmillan. https://doi.org/10.1057/9781137027832.

Rogers, C. R. (1961). *On becoming a person: A psychotherapists view of psychotherapy*. Boston: Houghton Mifflin.

Sewell, W. H. (1992). A theory of structure: Duality, agency, and transformation. *American Journal of Sociology, 98*, 1–29. https://doi.org/10.1086/229967.

Springer, S., Birch, K., & MacLeavy, J. (Eds.). (2016). *The handbook of neoliberalism*. New York: Routledge.

Zeidler, D. L. (2016). STEM education: A deficit framework for the twenty first century?: Sociocultural socioscientific response. *Cultural Studies of Science Education, 11*, 11–26. https://doi.org/10.1007/s11422-014-9578-z.

Darren Hoeg is an Assistant Professor of secondary science and environmental education at the University of Alberta. His work is centered on development of Science, Technology, Society and Environments (STSE) curriculum, teaching and learning. His researched focuses on examining sociocultural productions of nature, particularly in school science discourse, how this is involved in constituting human relationships with nature, and how sustainable relationships with/in nature may be initiated through alternative school science approaches.

Larry Bencze is an Associate Professor in Science Education at the Ontario Institute for Studies in Education, University of Toronto. His teaching and research emphasize history, philosophy and sociology of science and technology, along with student-led research-informed and negotiated socio-political actions to address personal, social and environmental issues associated with fields of science and technology.

Sarah El Halwany is a PhD candidate at the Ontario Institute for Studies in Education (OISE), University of Toronto. She finished her bachelor of science and her master's in science education from the American University of Beirut, Lebanon. Her research interests are on promoting meaningful learning/teaching through socio-scientific issues and on role of emotions in science education.

Erin Sperling is a doctoral candidate at OISE/UT with a research focus on ecojustice education relating to food literacy through qualitative and community-based practices. She is also a sessional lecturer in Science Methods and Environmental Education. Her recent publications include articles in Green Teacher magazine, Environmental Education Research and the Canadian Journal of Science, Mathematics and Technology Education.

Majd Zouda is a Ph.D. candidate in science education at Ontario Institute for Studies in Education (OISE), University of Toronto. Her doctoral thesis focuses on STEM education in Canadian secondary schools. Research experience includes youth socio-political activism on socioscientific issues, and peace-building citizenship learning. She has an M.Sc. in medical microbiology, and has worked as a middle-school science teacher in Cambridge-certified Pakistan International School of Damascus (PISOD), Syria. She has recently been awarded the Social Sciences and Humanities Research Council of Canada (SSHRC) doctoral fellowship award.

Chapter 14
A Critical Co/Autoethnographic Exploration of Self: Becoming Science Education Researchers in Diverse Cultural and Linguistic Landscapes

Jennifer C. Park and Sara E. D. Wilmes

14.1 Globalization Influences Our Lives, and Thus Influences Our Research

In today's globalized society, linguistic and cultural worlds collide bringing people, cultures, and languages together in diverse ways that can influence a person's identity and sense of self. Due to the porous boundaries of people's identities, increased globalization can lead to identity confusion, which can influence how open individuals are to integration (Hermans and Hermans-Konopka 2010). This confusion increases as countries across the world experience what Vertovec (2006) describes as super-diversity, a "world in one city" (Benedictus and Godwin 2005, p. 2). Super-diversity can be described as a dynamic interplay of variables among increasing numbers of new, small and scattered, multiple-origin, transnationally connected, socio-economically differentiated, and legally stratified number of immigrants throughout the world (Vertovec 2007). While some researchers have claimed that globalization can result in a loss of cultural diversity (Tomlinson 2003), the integration of plurilingual and pluricultural people in diverse contexts can also result in greater awareness of diversity. To understand the impact of globalization on an individual's identity and sense of self, research can be a powerful tool. In particular, research that critically focuses on examining oneself can reveal new knowledge of the self that may inform one's research endeavors.

In this chapter, we share the process of collaborative autoethnography (co-autoethnography), that we used to individually and collaboratively examine and reflect

J. C. Park (✉)
University of Southern California, Los Angeles, CA, USA

S. E. D. Wilmes
University of Luxembourg, Luxembourg City, Luxembourg
e-mail: sara.wilmes@uni.lu

J. Bazzul, C. Siry (eds.), *Critical Voices in Science Education Research*,
Cultural Studies of Science Education 17,
https://doi.org/10.1007/978-3-319-99990-6_14

upon our experiences at the start of our doctoral research paths. We used this critical methodology to help us make sense of our individual and collective experiences in two different multilingual/multicultural research settings – South Korea and Luxembourg. Specifically, we attempt to tease apart the relationship that exists between language, culture, and self. In addition, building from a discussion of the benefits and challenges of implementing this method, we share what we learned from this narrative process about ourselves, and about how we engaged and continued to engage in science education research. We share our co-autoethnographic process that helped us push the boundaries of our research, and that gave each of us a deeper understanding of our own positionality within our research contexts. We do so with the hopes of encouraging other researchers to undergo such processes in order to further investigate their own self and the position of their selves in their research.

14.2 Our Study of Ourselves

This study shares our narratives that are based on our experiences as researchers and science educators in a globalized world. Our co-autoethnographies and analysis are twofold and centered around critically exploring what it means to be researchers who are language and culture learners, or *newcomers*. By newcomers, we mean *us*, as we each arrived in a culture that was new to us when we started our own respective doctoral programs. Specifically, we explored these research questions: *How am I positioned as a language learner in this new multicultural/multilingual context? What impact does this have on my work as a science education researcher, and in academia, in general? How does my experience reflect that of students in multilingual/multicultural science classrooms?* In the sections that follow, we elaborate our co-autoethnographic process of our time as *newcomers* and how through the lens of positionality and intersectionality we were able to tease apart our relationship to language, culture, and self.

14.2.1 What Is Collaborative Autoethnography?

Autoethnography is a form of qualitative research that involves systematically looking at the self and the social phenomena involving the self (Ali-Kahn 2011). This method focuses on data that is collected, analyzed, and interpreted based on a researcher's understanding of the world and the people around them. This type of research is context conscious and has deliberate intentions of connecting the self with others, the self with the social, and the self with surrounding contexts (Wolcott 2004). The co-autoethnographic approach adds an additional layer to autoethnography, in that it involves processes of sharing autoethnographies in a community with others as an additional layer(s) of analysis and critical reflection.

Co-autoethnography as a research methodology is a powerful tool that uncovers data that can richly inform individuals, the community, and society as a whole. In adopting this research approach, it was key for us to understand it both as a theory and as a method. We briefly describe each in the next sections.

14.2.2 Autoethnography as a Theory

Autoethnography involves an iterative study of the *self* (auto) and its connections to *culture* (ethno) through systematic *analysis* (graphy) (Ellis and Bochner 2000). It is a research methodology that, as Sarah Wall (2008) explains, provides a way to give voice to personal experiences, and at the same time to advance sociological understanding. We adopted the stance of Coia and Taylor (2009, p. 2) who explain that, "we can never understand our own practice until we have some measure of understanding of our place in the execution of that practice". In other words, all circumstances and choices that we make, as well as the perceptions we adopt and develop, involve our beliefs, values, and prior experiences (Coia and Taylor 2009). Adopting this stance in our research provides us – teachers, researchers, individuals, etc. – the opportunities to reflect upon past experiences and the roles we undertook in various circumstances and to further explore connections between these past experiences and the way we perceive and understand things today. To this end, autoethnography as a theory emphasizes the importance of exploring and recognizing our own practices and assists us in unpacking the reasons for enacting certain practices in our *newcomer* situations.

14.2.3 Autoethnography as a Method

As Ellis et al. (2011) explain, when researchers use autoethnography they selectively write about past events in which they took part, then analytically look at these events to derive cultural meanings and connections. Building on this, we approached our autoethnographies by writing *evocative personal narratives* (Ellis and Bochner 2000) centered on our professional and personal experiences in our new multicultural/multilingual contexts; this involved our day-to-day and life-changing experiences related to culture and language. By focusing on our diverse professional and private lived experiences, we were able to reconstruct vivid textured narratives to critically analyze the factors that informed the development of our *self*. This included examining how we were positioned, how we positioned ourselves based on others' positioning, and how this positioning influenced the way we viewed and conducted science education research as language-learners in our multilingual/multicultural contexts. The autoethnographic narratives we constructed allowed us to identify and highlight key instances when our current understanding of self, cultural and linguistic structures, and our attitude/response to being multicultural and multilingual newcomers was evident. Then, working together to analyze our narratives and perspectives allowed us to serve as critical peers. In this way, co-autoethnography incorporates multivocality into the examination of one's self. (Hernandez et al. 2017). This process afforded us opportunities to further investigate our "self" and our perspectives in both our professional and personal experiences.

14.2.4 Collaboratively Writing Autoethnography

The process of collaboratively gathering and analyzing data based on personal narratives of our lived experiences initially seemed difficult and impossible. Would we be safe expressing our true voices? How critically do we want to explore the past? How can we learn from past experiences and share what we learned without implicating those close to us who might have been involved? Yet, the co-construction of themes and ideas, as well as the challenging task of identifying the self – by listening to others' varying perspectives – bears richer data than that which emanates from a solo-researcher's narrative. According to Ngunjiri et al. (2010), collaboration between researchers provides space to stir one another's memory, to probe questions unsettling to one another's assumptions, and to challenge one another for greater detail through constructive discussions. The collaborative discussion about and sharing of narratives helps focus each the of the researchers to mutually be accountable for one another's writings and analysis (Ngunjiri et al. 2010). To this end, we worked past our initial doubts and began individually writing our exploratory narratives. We then began collaboratively analyzing our understanding of our self/identities. We chose to examine our identities on two planes: First, in each of our native contexts versus in our *newcomer* multicultural/multilingual contexts; and second, our awareness of and connections between the language and culture and its connection to our research. The next section describes the transition we made from what we knew as our "norm" to being positioned as a *newcomer.*

14.3 Our New Multilingual, Multicultural Worlds

Prior to this joint venture in research, we each had lived most of our lives in a similar setting – speaking English while living in the United States. Then, as we began our journey as science education researchers, we each dove into a new context that was linguistically and culturally very different and much more diverse than we imagined. Learning to live in these new multilingual and multicultural contexts brought challenges that forced us to confront who we were, or thought we were. We were faced with having to think about the ways others perceived us, as well as how we perceived ourselves – in our *old* and *new* surroundings – and how this positioned us to think and act in certain ways. Thus, this research approach assisted us through a process of collaborative analysis, in shedding light on various themes and tensions involved with our becoming researchers in new contexts and our positions as *newcomers*.

14.4 Theoretical Approaches

In this next section, we elaborate two theoretical lenses that grounded our co-autoethnographic exploration of our *newcomer* selves. Our combined use of these lenses afforded us opportunities to connect what we were experiencing to what has been discussed in prior literature on the intersection of language and identity, and to extend this to our own lived experiences as newcomers.

14.4.1 Positionality

Positionality is generally revealed through discursive practices that result in the establishment of the self and others. It also simultaneously can serve as a resource through which all persons involved can negotiate new positions (Harré and Langenhove 1991). Positionality does not have an end product since positioning and re-positioning is a continuous, ongoing process that takes places not only in relation to other people, but also in relation to oneself (Hermans and Hermans-Konopka 2010). Thus, it involves an ongoing production and reproduction of the self (Davies and Harré 1990).

The lens of positionality highlights the concept of "others" and "othering", which is not outside of the self, but rather an intrinsic part of the self (i.e. self-conflicts, self-criticism, and self-agreements) (Hermans and Hermans-Konopka 2010). Since one's position gradually and continuously changes, it can be an important factor to consider when reflecting upon the self. A position can be influenced both by the way one is positioned by surrounding communities and by society as a whole, which in turn influences how one intrinsically positions oneself. The acknowledgement of the processes involved in positioning and how influential surrounding communities can play a critical role in the positionality of researchers and their work is important. In any research, we believe it is essential for researchers to recognize and critically examine their identity. Since positionality does not occur in isolation, the cultural, social, historical, and linguistic contexts that surrounded us were essential to consider. Thus, in addition to positionality, we drew upon intersectionality as a theoretical lens, which helped us to consider the complexity of our multicultural/multilingual contexts, yet to provide a clearer understanding of their complexity.

14.4.2 Intersectionality

As we began thinking about the different positionalities involved in the development of our selves, and our identities, it helped us to additionally apply the lens of intersectionality. This lens can serve as a powerful tool to examine and bring awareness to crucial factors that intersect in the shaping of our identities, as individuals and as researchers (Martin et al. 2013). Employing intersectionality as a research lens helped us identify and problematize consequences that resulted from simultaneously interacting factors in our lives related to our genders, our positions as researchers, and our positioning relative to the diverse linguistic and cultural factors we were facing as new researchers and newcomers in our communities. The intersecting factors that we identified and distilled from our discussions surrounding our written narratives were relative to each of our distinct ethnic backgrounds, nationalities, language resources, and our past experiences with education (both our own, and also as we each worked as science educators prior to starting our doctoral studies), and being women in the sciences. Identifying these intersecting factors allowed us the space to construct and deconstruct ourselves and to examine how these facets informed our identities – who we were, who we are, and who we are striving to become as researchers in science education. Applying the lenses of positionality and intersectionality throughout our co-autoethnographic approach equipped us to carefully interrogate each of our beliefs, thoughts, practices, and most importantly our mindsets in relation to our self and to others in our own respective research contexts.

14.5 Method of Collaboration

In this section, we discuss the process we used to engage in collaborative autoethnography. Specifically, we offer examples of how we generated and shared individual narratives that described our experiences as newcomers to new countries, and as novice science education researchers. We first met at a Cultural Studies of Science Education (CSSE) writing conference in Luxembourg in the summer of 2013. We both attended a workshop on education in multilingual and multicultural contexts and were introduced to each other, specifically since we were both just beginning our doctoral studies. As we talked, we realized that though we were living in very different contexts, we had a lot in common, which initiated our autoethnographic collaboration. Once we informed ourselves about the theory and methodology involved in participating in a co-autoethnographic process, we began by discussing ethical issues related to our process. Ellis and Bochner (2000) provide an excellent guide for considering the particular ethical issue involved in conducting autoethnography. The two main issues we discussed included, assuring each other that all information shared with one another would not be shared elsewhere without permission, and that publications would maintain the anonymity of those involved unless they provided consent.

14.5.1 Setting Goals for Our Narrative Writing

Once we discussed and agreed upon ethical guidelines, we started writing individual narratives to explore our backgrounds of who we are, where we initially lived, how we came to our respective new multilingual/multicultural context, what we were interested in pursuing/studying in this different context, and how/why this experience was important for us as researchers in science education. We began each of our narratives by detailing our own cultural, linguistic, and social backgrounds, making sure to describe our new contexts in which we each lived and worked. We provide a summary of who we are in Table 14.1.

At the start of our collaboration, we discussed the challenges we faced as we left what we once knew and embarked on living and conducting research in culturally and linguistically unfamiliar places. We decided it would be interesting to share our experiences first before diving into the various positions and intersecting factors that were shaping our identities. We provide a short narrative of our elves below that served as a basis for our initial collaborate analysis.

14.5.2 Narrative Summaries: A Look at Our Self: What We Knew and Choosing to Leave It

Jennifer I was born and raised in the United States and lived as a language minority for most of my life, always struggling to figure out where and how I belonged. I was limited in both the English and Korean languages; neither was foundationally set when I was growing up. I only spoke and verbally understood Korean, while academic English was a constant challenge for me to overcome. My transition to Korea

Table 14.1 Brief summary of our backgrounds that provided the initial context of our co-autoethnographic study

Jennifer	Sara
Born and raised in the United States as an ethnic Korean	Born and raised in the United States as an English-speaking Caucasian
Languages: L1 – Korean, L2 – English	Language: English
Struggled with academic English for years	Learned "foreign" languages in school
Transition	*Transition*
Relocated to South Korea in 2011 for science education doctoral program	Relocated to Luxembourg in 2010
Identity as a Korean-American, *Jaemi Kyopo*	Knew none of the languages (Luxembourgish, German, French)

was not easy. Regardless of being ethnically Korean, I realized that the language tools I came with were "outdated"; in other words, I was speaking Korean from the 1980s – the language my immigrant parents spoke after leaving the rapidly changing country. Upon my arrival, I was quickly identified as a *Kyopo* (Korean-American) who did not speak or understand Korean well. My relationship to language jumped backwards; my Korean skills resembled my limited English language skills that I struggled with back in the United States. I was a language learner once again. I felt as though I was never able to "win". My perspective started shifting as I began talking and collaborating with Sara. I started to recognize the benefits of my struggles and how they can inform my future years on both a personal level and towards research – generating opportunities to connect with the greater science education community.

Sara My move to Luxembourg in 2010 was the first time I took a critical look at my relationship to the languages I spoke. I was living in a new country and spoke none of the three languages (Luxembourgish, German, or French) at a level that would allow me to communicate with people in day-to-day interactions, such as at the grocery store, or to convey to a doctor why I was not feeling well. I was in many ways, for the first time, silenced. To me growing up as a Caucasian female who was raised in a middle-class family in the United States speaking English, knowing a foreign language was something exotic. It was a skill that one used on trips and in other-away lands. I describe it this way purposefully to illustrate the fairy-tale view of my relationship to language. When I moved to multilingual Luxembourg, the glass bubble shattered. I could not express myself. I could not connect. I was placed in migrant classes where people assumed I did not know how to read. I began to question everything through the lens of language. It was at this point that I met Jennifer and we decided it could be interesting to explore each of our experiences in our new contexts.

14.6 Collaborative and Recursive Data Collection and Analysis

The process of co-autoethnography is a backwards, forwards, and sideways movement that involves individual narratives (vignettes) and responses that have been written into them through talking, theorizing, and analyzing contexts with other collaborating members (Coia and Taylor 2009). To this end, we conducted multiple recursive cycles of analysis. Since we were living in different countries, we utilized as much technology as possible to assist us. We exchanged narratives via email and had collaborative conversations via Skype. We utilized a Skype plug-in to record our video conversations, which allowed us to collect data in several different layers. Our data sources were comprised of our narrative writings, email exchanges, and video recordings of our real-time conversations. We further explain this process in the sections that follow.

14.6.1 Generating Narratives

The key to the data that we collected was our intentionality to hold fast to an ethnographic approach. Writing down our narratives took precedence over our verbal conversations in order to provide space for us to individually reflect upon our self and our experiences. According to Faith Ngunjiri et al. (2010), dialogue is richest when method takes precedence over personality, thus our writings were vignettes that explored themes that we separately reflected upon in writing first.

In order to critically look at ourselves as both language and culture learners, our vignettes were focused on answering the following questions: *What is happening with language in my life right now? What language spaces do you encounter in your current context? What happens in those spaces?* Although we started by addressing these questions, we also freely wrote other feelings and/or thoughts that were relevant (see Table 14.2) and that arose while writing. Whether our experiences were in the past or the present, at a younger age or something experienced recently, we created a plethora of narratives through this free-writing process.

Table 14.2 Example of the thoughts we initially noted, individually, in efforts to begin writing our autoethnographic narratives

Jennifer	Sara
Tendencies in Korea – Wanting to speak the hybrid language, *Konglish* (Korean and English hybrid language) – Continually surrounding myself with English-speaking people (mostly the expat community) – Unknowingly encouraging native Korean(s) speakers to use *Konglish* by my use of the hybrid language with them; *Konglish* was only used when both languages are easily recognizable for both me and the native Korean(s); they began to realize my comfort in speaking out words that are difficult in Korean	How does being a language learner affect my work as a PhD student? – I delay writing emails – I draft more drafts than I would in English – I don't understand everything – I see more than I hear – I pick up on subtle clues because I can't rely on the spoken – I miss jokes and cultural generalities

14.6.2 Analysis of Our Narratives

The narratives that we wrote followed the model shown in Fig. 14.1. We followed a recursive cycle that involved individually writing, discussing the writing, teasing apart the obvious and hidden pieces of useable writing, and re-writing to improve or add new content. The vignettes were exchanged over several cycles so that we could read one another's stories, try to understand where the other person was coming from, probe more in-depth with questions to further one another's thought processes, and to help support claims with literature-based resources.

After the first round of exchanging vignettes, we utilized the Skype platform to begin discussing and our ideas, which mostly involved checking in with one another, further describing areas we found interesting or needing more explanation, and identifying themes from what we shared. The entire Skype conversation was recorded using a plug-in and we produced transcripts of our conversations. This data, which we analyzed individually and collaboratively, was used to identify themes that emerged from both our narratives and conversations together about our (Fig. 14.1) and to recognize if ideas/thoughts we wrote from what we recollected and the things we said were aligned within each of our stories and retellings (see Fig. 14.2).

Next, we individually conducted transcript review from the first round of analysis. This prompted us to come together again to write detailed explanations to questions that arose during our first round of narratives and Skype transcript analyses.

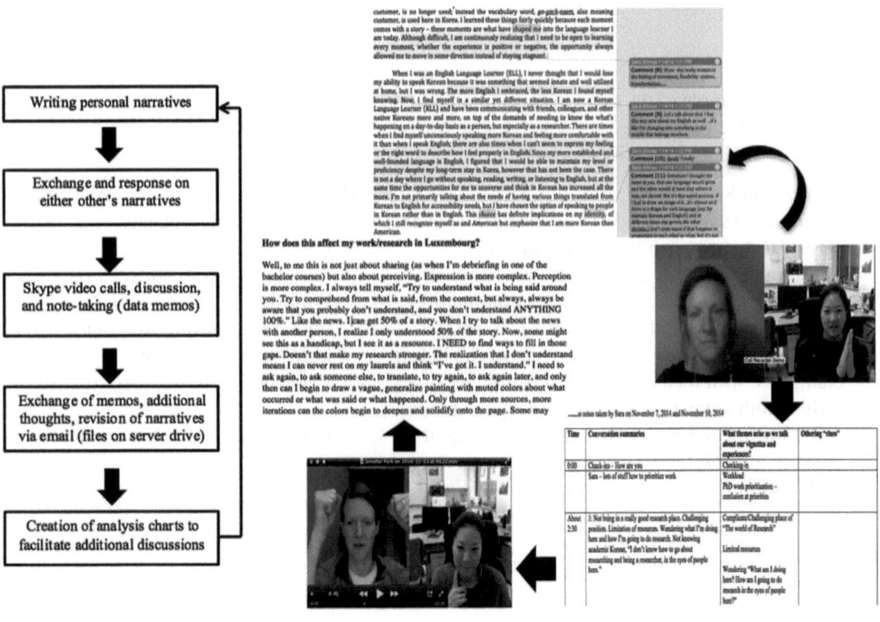

Fig. 14.1 A flow chart of our collaborative recursive autoethnographic process

Skype Memo of 20141105 Skype Jen and Sara first Autoethno Conversation

Meeting with Jen on 5/11/14 to exchange ideas comments and talk about the first round of work that we exchanged the week prior.

Memo notes taken by Sara on November 7, 2014 and November 10, 2014

Time	Conversation summaries	What themes arise as we talk about our vignettes and experiences?	Othering "clues"
0:00	Check-ins – How are you	Checking-in	
	Sara – lots of stuff how to prioritize work	Workload PhD work prioritization – confusion at priorities	
About 2:30	J: Not being in a really good research place. Challenging position. Limitation of resources. Wondering what I'm doing here and how I'm going to do research. Not knowing academic Korean, "I don't know how to go about researching and being a researcher, in the eyes of people here."	Complicate/Challenging place of "The world of Research" Limited resources Wondering "What am I doing here? How am I going to do research in the eyes of people here?"	
	S: Not being able to get a grant because I don't know the local languages. How we did data collection in a team to overcome the language barrier.	Strategies to make research work when one is researching in a language setting in which they are not "fluent".	3:13 THEY are like, YOU can't do research here, YOU don't know the languages. (How do researchers clue into positionality what techniques do they use?)

Fig. 14.2 Sample analytic memo constructed while listening to the video data and reading Skype transcripts

This led us back to the drawing board to recollect additional experiences related to the themes we extracted, and to develop additional vignettes. Our recursive approach was our attempt to analyze existing data, while adding on layers of new data through a collaborative, reflective, and participatory process involving both of us (Coia and Taylor 2009). This collaborative and recursive process uncovered very interesting themes and ideas that we did not necessarily anticipate, and that would not have been possible to uncover if we had done this process on our own, in our own respective contexts.

14.7 Insights from Our Analyzed Lived Experiences

As a result of collaboratively analyzing our narratives, we discovered insights into the complex themes and challenges that arise when one attempts to live, work, and conduct science education research in new cultural and linguistic landscapes. While each of our positions and intersecting factors were unique to our own contexts, this collaborative investigation allowed us to draw comparisons that helped us to critically compare our situations. The insights from this collaborative process supports current theoretical understandings regarding the positioning of newcomers in multilingual/multicultural learning environments. It additionally helped us to realize the methodological strength of co-autoethnography as a way to explore our positions and the multitude of intersections present in our everyday experiences.

During the analysis of our narratives and Skype transcripts, we made sure to examine the data resources using the lens of positionality and intersectionality. In this way, we became more conscious and intentional about identifying factors that positioned us, as well as how we positioned ourselves, in certain ways within our personal and research contexts. We identified interesting terms and phrases that were not necessarily written, but that we repeatedly verbalized during our conversations. The next sections elaborate realizations and insights that emerged during our collaborative analysis.

14.7.1 We Were Often Positioned as the "Other"

In analyzing our Skype discussions about our narratives, it became apparent that in our speech and descriptions of ourselves and our contexts, that we both embraced the category of the *other*. By this we mean, we positioned ourselves as not as belonging, but as being the *other (newcomer)*. During our first Skype conversation, the terms "us" and "them" were prominent. We were alarmed that we used these words frequently when describing our lived experiences as newcomer PhD students in multilingual/multicultural contexts. For example, there was a complex interplay of positionality, culture, intersecting factors – being female, foreigner PhD students, and language as revealed in the following excerpt we extracted from our analysis:

> …from my lab, it's [daily tasks and behavior] very fluid because I know the culture and we all speak English in here, but when I walk down the hallway I don't know who's gonna speak what to me…and I should speak Korean, it's fixed, the culture is Korean. I have to bow to professors [and other colleagues] when I see them or I have a choice to be out of that fixed context, putting myself as the other by saying, "hi" instead of bowing.

Our analysis supports prior findings that explain that despite increasing linguistic and cultural diversity in parts of the world, there are *norms* and *boundaries* that still exist in many communities that maintain division. These often lead to the creation of marginalized groups that are positioned as the *other*, a group of people separated from communities due to systemized categories. In our case, through our collaborative process, we found these norms and boundaries not only in the communities we were participating in, but also in ourselves, in the ways we spoke and described our own abilities and interactions. If we relate this to the *othering* and marginalization of culturally and linguistically diverse students in science classrooms, it is clear that we, as is the case for many of the students with whom we conduct research, were often positioned as the *other*, and positioned ourselves as the *other*, which had a direct impact on our lives and our research.

We, too, conveyed a deficit view of multilingualism (language proficiency). It became clear that in the stories that we shared, reflected upon, and analyzed, we were almost always portrayed by others, or by ourselves, as deficient in both our new languages and cultural settings (countries and in academia). We did not honor the strengths that we had developed during the critical period of coming-to-be-proficient in a new cultural and linguistic landscape and our less-than-fluent abilities. It may have been our automatic instinct of trying to "fit in" and acculturate to the multicultural/multilingual context, but at the same time the level of value our surroundings placed on the diversity of culture and language we claimed played a crucial role. For example, we saw ourselves as not fluent in our new languages, and thus, not being good enough. We were unconsciously positioning ourselves, and found instances in our narratives of others positioning us as well, as deficient relative to assumed monolingual norms. Our analysis revealed the unconscious ways we too were positioning ourselves as deficient. Revealing the ways we were self-positioning provided us the opportunity to be critical of the unconscious positioning of ourselves and how this intersected with our roles as educators and researchers. It also allowed us the space to consider how we might project these views onto our students and in our research.

14.8 Collaborative Autoethnography Illuminated Our Positions and Subjectivities

Our collaborative autoethnography study involved a reflexive and recursive processing approach to investigate the way we position ourselves in the stories we tell and how circumstances are internalized as part of our identities; not taking such steps to break down our narratives would obscure the understanding of our positioning. Having a better understanding of yourself, personally and as an educational researcher, is critical since we hold "our own frames of reference" which surface during the analysis phase of research (Erickson 1986) (in Glesne and Peshkin 1992); this can cause biased results making the data irrelevant and invalid. Collaboratively utilizing this method afforded us the ability to engage in a mutually supportive practice of voicing and sharing our difficult positions and frames of reference as language learners in new contexts. Co-autoethnography provided a mechanism for us to create community around personal stories and histories filled with feelings of being "othered", positioned "outside" of a culture thus acting upon the positioning characteristics, and being "deficient" in a language or knowledge in a culture. In addition, this approach allowed us to retell our stories in positive and non-deficit ways, which empowered us to re-position ourselves in our respective multicultural/multilingual contexts.

Collaborative autoethnography helped us form a space in which we could look at our own selves with a critical gaze through supportive collaboration with each other. In the space we created through our use of this methodology, we gained back parts of our voices we had lost as newcomers. It also allowed us to critically examine these voices within safe places as we moved forward into academia, where critical spaces were not created within the official structures of our doctoral programs. Our collaboration on this research project was a time for us to pause and reflect on our journeys. In doing so, we were able to "pause and critically tease apart (our) positions in order to breathe new life into our theorizing and our writings in all its different forms) (Giampapa and Lamoureux 2011, p.129). It afforded us the space to be critical of who were becoming as newcomers to academia, and to examine the voices we were forming within our new roles.

14.9 Implications for Science Education Researchers and Educators

This study has implications for science educators and researchers in multicultural/multilingual contexts as it emphasizes the importance of bringing to light researchers' and teachers' predispositions toward language learners in general, and in research contexts in particular. As the writer and activist Barbara Deming (n.d.) once explained, "the longer we listen to one another – with real attention – the more commonality we will find in all our lives. That is, if we are careful to exchange with one another life stories and not simply opinions." The collaborative process of listening to ourselves and listening to each other helped us understand the power of this personal, yet collaborative, process. We created a safe space for us to assume a critically reflexive stance towards our own lives and experiences while in the process of becoming researchers. It enabled us to change how we positioned ourselves and others, both in our daily interactions and in our work with research participants. We offer our example of the use of co-autoethnography, as a way to explore one's position as both an individual and a researcher, in that it can bring light to one's own perspectives in diverse cultural and linguistic contexts.

References

Ali-Kahn, C. (2011). Shaken and stirred: On coming to critical praxis. In T. Kress (Ed.), *Critical praxis research: Breathing new life into research methods for teachers*. Dordrecht: Springer. https://doi.org/10.1007/978-94-007-1790-9_11.

Benedictus, L., & Godwin, M. (2005 January 21). Every race, colour, nation, and religion on earth, The *Guardian*. Retrieved from https://www.theguardian.com/uk/2005/jan/21/britishidentity1.

Coia, L., & Taylor, M. (2009). Co/autoethnography: Exploring our teaching selves collaboratively. In L. Fitzgerald, M. Heston, & D. Tidwell (Eds.), *Research methods for the self-study of practice* (pp. 3–16). Netherlands: Springer. https://doi.org/10.1007/978-1-4020-9514-6_1.

Davies, B., & Harré, R. (1990). Positioning: The discursive production of selves. *Journal for the Theory of Social Behaviour, 20*(1), 43–63.

Deming, B. (n.d.). BrainyQuote.com. Retrieved June 1, 2017, from BrainyQuote.com Web site: https://www.brainyquote.com/quotes/quotes/b/barbaradem325862.html.

Ellis, C., & Bochner, A. P. (2000). Autoethnography, personal narrative, reflexivity: Researcher as subject. In N. K. Denzin & Y. S. Lincoln (Eds.), *The handbook of qualitative research* (2nd ed., pp. 733–768). Newbury Park: Sage.

Ellis, C., Adams, T. E., & Bochner, A. P. (2011). Autoethnography: An overview. *Historical Social Research/Historische Sozialforschung, 36*, 273–290.

Erickson, F. (1986). Qualitative methods in research on teaching. In M. C. Wittrock (Ed.), *Handbook of research on teaching* (3rd ed., pp. 119–161). New York: Macmillian.

Giampapa, F., & Lamoureux, S. A. (2011). Voices from the field: Identity, language, and power in multilingual research settings. *Journal of Language, Identity, and Education, 10*, 127–131. https://doi.org/10.1080/15348458.2011.585301.

Glesne, C., & Peshkin, A. (1992). *Becoming qualitative researchers: An introduction*. New York: Longman.

Harré, R., & Langenhove, L. V. (1991). Varieties of positioning. *Journal for the Theory of Social Behaviour, 21*(4), 393–407. https://doi.org/10.1111/j.1468-5914.1991.tb00203.x.
Hermans, H., & Hermans-Konopka, A. (2010). *Dialogical self theory: Positioning and counter-positioning in a globalizing society*. Cambridge: Cambridge University Press. https://doi.org/10.1017/CBO9780511712142.
Hernandez, K.-A. C., Chang, H., & Ngunjiri, F. W. (2017). Collaborative autoethnography multivocal, relational, and democratic research: Opportunities, challenges, and aspirations. *Auto/Biography Studies, 32*(2), 251–254. https://doi.org/10.1080/08989575.2017.1288892.
Martin, S., Wassell, B., & Scantlebury, K. (2013). Frameworks for examining the intersections of race, ethnicity, class and gender: An analysis of research on English language learners in K-12 science education. In J. A. Bianchini, V. A. Akerson, A. Calabrese Barton, O. Lee, & A. J. Rodriguez (Eds.), *Moving the equity agenda forward: Equity research, practice, and policy in science education* (pp. 81–98). Dordrecht: Springer.
Ngunjiri, F. W., Hernandez, K. C., & Chang, H. (2010). Living autoethnography: Connecting life and research. *Journal of Research Practice, 6*(1), 1. https://doi.org/10.1080/08989575.2017.1288892.
Tomlinson, J. (2003). Globalization and cultural identity. *The global transformations reader, 2*, 269–277.
Vertovec, S. (2006). The emergence of super-diversity in Britain. *ESRC Centre on Migration, Policy, and Society.* Working paper No. 25 University of Oxford, 1–44.
Vertovec, S. (2007). Super-diversity and its implications. *Ethnic and Racial Studies, 30*(6), 1024–1054. https://doi.org/10.1080/01419870701599465.
Wall, S. (2008). Easier said than done: Writing an autoethnography. *International Journal of Qualitative Methods, 7*(1), 38–53. https://doi.org/10.1177/160940690800700103.
Wolcott, H. F. (2004). The ethnographic autobiography. *Auto/Biography, 12*(2), 93. https://doi.org/10.1191/0967550704ab004oa.

Jennifer C. Park is currently a lecturer at the Rossier School of Education, University of Southern California in Los Angeles, California. She is working in the MAT-TESOL department focusing on effectives ways to instruct, engage, support, assess, and provide learning opportunities for English language learners through varying sociocultural lenses. Her doctoral research at Seoul National University examined the teaching and learning of K-12 multicultural students in Korean classrooms, in addition to seeking ways to improve Korean teachers' professional development/training via action research, cogenerative dialogue, and other methods of qualitative research.

Sara E.D. Wilmes is a researcher at the University of Luxembourg. She currently works with the SciTeach Project (PI: Christina Siry) to develop a collaborative teacher education network for sustained changes in primary science education in Luxembourg. Her doctoral research examined the use of student-driven inquiry-based science instruction using an instructional and professional development approach she designed to address the needs of cultural and linguistically diverse primary school students in Luxembourg. Her research explores science education in multilingual contexts and the use of postmodern research methodologies to tease apart intersections between science learning and diverse communicative resource use.

Chapter 15
Resistance to Divergent, Child-Centered Scientific Inquiry in the Elementary School and at the University: An Autoethnography of a Science Educator

Brian Stone

15.1 The Philosophical, and Socio-professional Context of a Critical Science Educator

I explored the context of my research, teaching, and philosophy as a science educator through the mode of autoethnography due to the "aesthetic and evocative" nature of a research paradigm that questions the "ontological, epistemological, and axiological limitations" of more quantitative, "neutral" approaches to social science (Ellis et al. 2011). I consider myself an outsider. Though I have been in education for over a decade, I do not fit in with or promote the current predominant trends and paradigms of the dominant practices in education. I am a critical science educator, and question both the objectives and outcomes of the P-16 educative process, which often narrows the curriculum, limits academic freedom and democratic processes (French 2005), and devalues the inimitability of children.

In order to present my work as "self-consciously value-centered rather than pretending to be value free" (Bochner 1994), I composed the following story of my journey as a science educator in what I believe to be a hostile system, which has disregarded, marginalized, and ostracized child-centered pedagogy and philosophy. Additionally, I provide examples from my research and teaching that demonstrate a struggle with the dominant culture of education. Finally, I discuss pathways for creating discourse and challenging the status quo in science education.

B. Stone (✉)
Northern Arizona University, Flagstaff, AZ, USA
e-mail: brian.stone@nau.edu

J. Bazzul, C. Siry (eds.), *Critical Voices in Science Education Research*,
Cultural Studies of Science Education 17,
https://doi.org/10.1007/978-3-319-99990-6_15

15.2 A Guiding Metaphor for Change

In reflecting upon the context of my educational career, I was inspired with a metaphor for those like me trying to affect change in what I consider to be an adverse education system that values scores over people, and demands conformity, compliance, and homogenization. The metaphor is terraforming, or "earth shaping." The theoretical concept of terraforming has existed at least since 1942 and was coined by the science fiction novelist, Jack Williamson. According to NASA, "terraforming is the process of transforming a hostile environment into one suitable for human life" (NASA n.d.), and involves a series of changes to a planetary body and atmosphere to make it livable.

Mars is often the theoretical target of terraforming in our solar system, and it is indeed a harsh environment. However difficult and protracted a process it may be to terraform Mars, it is in some ways a greater challenge to transform our educational system to a workable, livable, healthy environment for all children and teachers. Yet, that is the challenge, for others, and myself to terraform our educational system. We live in an incredibly diverse, evolving, and organic society. Maya Angelou once said, "It is time for parents to teach young people early on that in diversity there is beauty and there is strength" (Michael 2012), and yet the system of education we use is sterile, rigid, and narrow. We often preclude different ways of thinking, and instead prefer, expect, and even mandate convergence of knowledge and skills. In doing so, we rank, order, and sort children, and many of those children are marginalized through standardized, systemic processes.

15.3 Paradigmatic Lens

Philosophically, I am a social-constructivist, described by Lisa Schreiber and Brielle Valle (2013) as knowledge that is constructed through one's various experiences, and is extensively shaped by social and cultural influences. Furthermore, I believe that children learn science through activity, tangible manipulations and discovery (Moyer et al. 2007), and that "the most engaging, relevant, and meaningful inquiry for children arises from their own questions, ideas, and interests" (Bloom 2006, p. 136). I would also consider myself a critical constructivist, defined by Taylor (1996) as an epistemology that concentrates on the socio-cultural constructions of the learner and serves the purpose of cultural transformation in various contexts. As a part of being a critical, social constructivist, I incorporate "culturally responsive science teaching," which caters learning to the individual needs of children with regard for their diverse cultural and life experiences (Wallace and Brand 2012).

I consider myself child-centered, which is defined as children being intrinsically engaged in their own education and constructing knowledge through their own unique experiences (Coughlin et al. 2000). In being child-centered, the teacher's role is to foster learning and provide for the interests of individual children, respect the unique strengths and needs of every child, promote the innate curiosity of the individual, and nurture collaboration (Coughlin et al. 2000). These perspectives are in direct opposition to curriculum-centered practice, defined as pedagogy that is "most effectively and efficiently transmitted through methods that impose curricular order and is characterized by pedagogical methods that presume teacher as authority, learning through repetition, and learning as a quantifiable outcome" (Pinnegar and Erickson 2010, p. 849–850).

15.4 Contrasting Views with the Dominant System

As an elementary teacher and university faculty member, I have seen a complex, dynamic spectrum of paradigmatic approaches to education. Most of those approaches fit within a dominant, normalized context that promote conformity and even oppression (Vinson et al. 2001). For example, many of my peers and colleagues over time have framed their courses from a curriculum-centered mindset, often with the same guiding question: "How do I get my students to conform to the standards?" I have heard many versions of this question from early childhood teachers through graduate-level professors, and the conversation inevitably centers on data, specifically, test scores at the elementary level, and outcomes in higher education. However, the data-driven rhetoric is contributing to underdeveloped, uninspired pedagogy that is lowering the quality of education as well as motivation and engagement across the profession (Vinson et al. 2001). Though the pedagogical approach may vary, the concern is always for convergent, homogenized outcomes, prescribed and predetermined for students by external stakeholders who hold stake not in the growth or learning of an individual, but in the mass data coming from testing. These processes have led to a simplification of education to the point that the institution has become a competitive machine in a capitalistic pursuit for satisfactory numbers (Vinson et al. 2001).

Martin Essex noticed the origins of this dominant trend all the way back in 1952 when he outlined the systematic assault on academic freedom by positing that teachers had become self-censoring in their fear, and that an emerging pattern at the time showed teachers losing their professional capacity to make decisions in their practice (Essex 1952). Essex observed and noted this trend of eroding academic freedom, noticing that very little protest came from increasing standardization and the degradation of professional judgment. Over decades, the erosion has affected multiple levels in education from elementary schools to universities.

As an elementary teacher, I experienced the pressure to conform, the lack of academic freedom, and organizational shifts towards a script. My approach strongly contrasts with a teacher-directed, structured inquiry approach, which is "insufficient for developing critical and scientific thinking and appropriate dispositions and attitudes" (Zion and Mendelovici 2012, p. 385). The standardized use of convergent inquiry would involve predetermined answers and follow a teacher-directed pathway that limits exploration and promotes reliance and dependence on the teacher to provide science questions, processes, and outcomes (Stone 2014).

Relating to academic freedom, and using the guiding metaphor of terraforming, I found that my approach, although informed by research, has been systematically rejected in the dominant atmosphere of a hostile world. The academic freedom to be different, to have an opposing viewpoint, to engage in critical research and scholarship is minimal, or in some cases, non-existent.

15.5 Pedagogy of a Critical Science Educator

Science education should involve open, authentic exploration, which "achieves a higher level of inquiry, in which the students become more familiar with the nature of scientific knowledge, develop greater inquiry skills and practices, and engage in higher-order thinking" (Zion and Mendelovici 2012, p. 385).

Some of the most valuable tools children have available are creativity and divergent thinking, which, in science, support discourse and often lead to open-ended, student-directed inquiry (Peters and Stout 2006). The individual development of scientific identity is also crucial. One scholar recommends "every student must seek and find their own academic and personal interests and construct an identity around them, which will sustain engagement and aid in their pursuit" (Kuhn 2007, p. 110). That fragile identity is tied to students' personal interests, which often have little value in the school system. So, in practice, child-centered education does not funnel the standards in a paced curriculum, delivered in a controlled, authoritative manner. Child-centered education respects the individual and promotes an organic, sophisticated exploration of meaningful, relevant, interesting content with the guidance of a teacher.

Despite my pedagogical approach and philosophical underpinnings, I never expected to find such substantial resistance from peers, administrators, university colleagues and even university students to child-centered, divergent, and diverse practices in science. In other words, the hostile environment of education has existed on the foundation of tradition, and is reinforced by a modernistic, industrial, positivistic, and behavioristic mindset that reduces curricula and instructional practice to a script, learning to a prescribed, sterile, uniform process, and assessment to discreet, isolated quantitative data. The system has maximized instructional and assessment efficiency while minimizing the role of the learner and the quality and divergence of the learning process. The system does not tolerate resistance or even different approaches on the part of the student and/or the teacher. When referring to this lack of freedom, one scholar states that the effect increases the "workload, [and] the demands of agendas imposed from above," which negatively affects climate through diminishing "confidence, creativity, and professional freedom" (Du Quesnay 2003, p. 49).

Albert Einstein once wrote, "It is, in fact, nothing short of a miracle that the modern methods of instruction have not yet entirely strangled the holy curiosity of inquiry; for this delicate little plant, aside from stimulation, stands mainly in need of freedom. Without this it goes to wrack and ruin without fail" (Einstein n.d.). Yet, the "holy curiosity of inquiry" does not belong to the individual; not the child, or even the teacher, but is derived from the textbook, the school curriculum, and the standards. The curriculum is influenced by interests other than meeting children's needs, and is often a dominant viewpoint that marginalizes certain groups. Linda Darling Hammond (2011) gives an indictment of this system:

> Meanwhile, the profession of teaching and our system of public education are under siege from another wave of scientific managers, who have forgotten that education is about opening minds to inquiry and imagination, not stuffing them like so many dead turkeys—that teaching is about enabling students to make sense of their experience, to use knowledge for their own ends, and to learn to learn, rather than to spend their childhoods bubbling in Scantron sheets to feed the voracious data banks that govern ever more decisions from the bowels of the bureaucracy. (para. 13)

Being a critical, child-centered educator in a system often hostile to child-centered methods and philosophy of teaching created a professional struggle, of which I have described briefly in the following sections.

15.6 The Culture of Compliance and Conformity

Sir Ken Robinson (2013) states that our educational model has built a "culture of compliance" and that teachers are coerced to "follow routine algorithms rather than to excite the power of imagination and curiosity" in students. Likewise, the expectations I faced from my elementary school and university leadership were in direct opposition to these goals. I felt the pressure to conform constantly.

15.7 The Beginnings of Resistance at the Elementary School

In my elementary school, I had a great deal of autonomy to run my class the way I wanted, at least when I started. I was privileged to teach during a relatively short window of time that allowed me to buck systematic trends in favor of child-centered practices like I described previously. Despite teaching during the middle of the NCLB reign, my school director asked me when I interviewed for the job how I felt about having complete autonomy as a teacher. I was not expected to follow a script, nor was I expected to use a certain textbook program or discipline program. When I started, I actually had academic freedom available in abundant supply, and I was treated as an expert and professional by my colleagues, the administration, and most parents. While others used professional funds for textbook programs, I bought science supplies and math manipulatives for my classroom. For a time, I was free, and I saw the great benefit of allowing teachers the freedom to exercise their professional capacity and judgment. To me, the healthy way to terraform is to bring the evidence, practice, and mentality to a small section of the hostile system and demonstrate how successful students and teachers can be. For a time, I was able to do just that in my school, as I know others have done in small pockets across the world.

As time progressed, the school leadership became increasingly obsessed with student scores on the state-mandated test. Our meetings changed in tone and were more data-oriented. A large sign was placed in the lobby of our school promoting the fact that we had become an "excelling school" as labeled by the state system, and showing that our number one priority had become test scores. Professional development opportunities chosen by the administration and school board were geared toward maintaining scores and boosting achievement on the test. I remember being called into the director's office on more than one occasion to specifically talk about my students' performance on the test. Interestingly, I used child-centered practices, and allowed for inquiry-based instruction in my classroom, and my students' scores were as good or often higher than my more traditional peers' scores. However, the growing data mentality of the administration and school board began to shape and dictate the processes of the school and even teachers' practice. The expressed expectations of the school leadership began to shift from an acceptance of child-centered practice to a critique despite high test scores and wide parental support.

While teaching, I was working on my doctoral degree, and a search opened at the local university for an elementary education generalist position. I applied and was offered the position. At the time, I felt the increasing pressures at the elementary level to perform, and the limitations on my professional capacity as teacher being imposed. I accepted the position. I viewed myself as a change agent, but had no idea how difficult change can be when dealing with an oppressive system steeped in decades of tradition.

The year after I left my elementary school, they adopted a standardized reading and math program. I had been able to integrate curriculum and provide equal time for subjects like science and social studies when I taught, but after I left, the emphasis was placed on math and language arts to the exclusion of other subjects. I know this because I worked in a Professional Development School model at the university, and placed my students for practicum experience in my former elementary school. I would ask my students what they were seeing as far as science instruction, and they would respond with "What science instruction?" The same was said for social studies. Upon hearing from my practicum students, my first thought was that I narrowly escaped such a constricting move by the school by moving to the university. I could see the hostile conditions of the system, and I had spoken with teachers at the school who I had known for years. Many did not like the new direction or the new program. They lamented the fact that it dictated their teaching schedules.

15.8 The Controlling Elementary School Context Through Narrative Research

My eyes were opened to the atrocities of the hegemonic system as I completed my dissertation research. I did a qualitative case study in two elementary classrooms that utilized inquiry-based approaches in science. However, a challenge for my research was to find teachers who used inquiry strategies. Most elementary teachers in the local district taught little to no science. The majority of their time was spent on math and language arts as a push to meet Common Core standards. Some would have their students participate in a school science fair and check off science like it was a completed subject. In informal conversations with district teachers as a part of my involvement with university practicum students, some teachers even mentioned the idea that the science fair was all the science instruction their students would receive. Other teachers alluded to the fact that science was more of an occasional subject, taught when teachers felt they had extra time, or felt guilty for not teaching science for a while. In defense of these teachers, many knew they needed more science instruction, but felt severely limited by external expectations, the school administration, standards, and the system at large. Some teachers would provide endless worksheets with simplistic memorization tasks and call it scientific inquiry, which was a misinterpretation of the term and its practice.

Eventually, I found two fourth-grade teachers at different schools in the district who used some inquiry instruction, and it was highly teacher-directed. I specifically was examining the impact of teacher-directed inquiry on students' use of their own authentic inquiries, and my research examined the impact on the climate of the classrooms with regard for students' interests and engagement. Through many hours of observation, student interviews, and science journal document analysis, I found that students were given zero time, resources, or materials to explore their own inquiries. Furthermore, students' own curiosities and lines of questioning had become inelastic, meaning that the more the teacher scripted the science process (generate science questions, processes, and outcomes) the less likely students were to come back to their original inquiries.

Finally, students had become disengaged, less interested, and reliant upon the teacher to "be the scientist." In other words, students did not see themselves as scientists, and they were dependent upon the teacher to give them answers, usually with little to no scientific process. This research confirmed other studies showing that students had become reliant and dependent upon the teacher to present science, and often were reluctant to utilize their own authentic inquiry because they had become conditioned to the teacher's scientific procedures (Haigh 1998). When asked about their expectations, the teachers reported having to follow the standards and the need to have convergent outcomes because their students would be tested on what was taught.

After I completed my dissertation research, I began focusing on the stories of my former students who were now teachers. After several years in higher education, I began to hear back from my students about the situations they faced in their teaching positions. As part of another study, I interviewed former students to create a narrative of how the system had built and now enforces compliance at all levels. One former student told me she "shouldn't have bothered with a college degree because the school controls every decision in the classroom including the color of folders students use." Due to a complete lack of autonomy, she mentioned that she was already burning out from teaching and that the job had made her feel "incompetent."

Other students had mentioned a lack of support from administration to even teach science and an overwhelming focus on math and language arts. Another mentioned, "Standards are treated like gods in our school." She told the story of how the school uses a very scripted textbook program and that she wanted to change the order of lessons for a specific concept. She asked the principal, who had to check with a representative from the textbook company. Ultimately, she was not allowed to change the order of the lessons, and mentioned that the students were placed at a disadvantage. Furthermore, my former student's school used a merit pay system, which they based on formal evaluations for each teacher. During the evaluations, the teacher had to read a script, and even the students had to ask certain questions. If the students said the right things, the teacher was given full points for her evaluation. Because merit pay was based on this type of evaluation, my student told me that teachers were coaching their students exactly what to say so they could earn full points. "We were like robots," she said.

Other students mentioned that their schools rigidly paced their curriculum so that all standards for all subject areas were expected to be taught by February of the academic year in order for the spring to be used for testing. Despite a desire to be child-centered, and provide engaging science activities that allowed for authentic inquiry, my students, now teachers were molded into curriculum-centered practice. For me, this was an increasingly difficult challenge, as I would have students report back to me that using inquiry, though strongly supported by research, was not encouraged in schools as they took too much emphasis away from core subjects. Also, inquiry took too much time and too many resources, therefore stifling the efficiency of the paced curriculum. I heard this often from students and district teachers.

Despite the dominant, authoritative trends, organizations like the National Science Teachers Association recommend more child-centered practice. In NSTA's (2010) position statement on "Principles of Professionalism for Science Educators" the following practices are recommended:

- Show respect for each individual and value his or her identity and cultural heritage;
- Recognize the abilities and strengths of students, as well as their unique learning needs;
- Model and emphasize the skills, attitudes, and values of scientific inquiry;
- Help students reflect as learners and use skills of inquiry to become effective problem solvers;

- Display and demand respect for diverse ideas, skills, and experiences of all students;
- Structure and facilitate ongoing formal and informal discussion based on a shared understanding of rules of scientific discourse; and
- Orchestrate discourse among students about scientific ideas

Additionally, the NSTA (2010) recommends that teachers have the "full support and active participation of school and community leaders." However, the realities I have experienced, and that my students have experienced, have been quite different. Little time or value is given to respecting students' individual needs, cultures, personal curiosity or discussion. The emphasis as described previously is on covering the standards in the most efficient way possible and securing the best possible scores for the school. Sir Ken Robinson (2013) states that our system has "de-professionalized" teachers. Alfie Kohn (1993) states that our system of control has increased the phenomenon of burnout among teachers and students. Diane Ravitch (2010) states the importance of academic freedom by suggesting the "essence of professionalism is autonomy, the freedom to make decisions based on one's knowledge and experience," but our schools have controlled teachers through "rewards and punishment for meeting targets" (p. 259).

The system is indeed hostile for someone who is a critical educator and believes in supporting the diversity, development and unique timeline for learning for every child. My students have not had the freedom as I once did. They have the choice to comply or leave. Even with extensive support systems, research, and strong professional practice, terraforming in the elementary school is a highly difficult endeavor, and resistance is met with systemic pushback.

15.9 Control in Higher Education

Though child-centered practices have been marginalized for some time at the elementary level, the university has had a shorter window of erosion. As I moved into higher education, I began to encounter a similar resistance to child-centered practice and divergent inquiry in science. Like my elementary school experience, I felt I had a great deal of autonomy and academic freedom at the university level when I began. However, a few years into my teaching, that began to change. Under our NCATE (National Council for Accreditation of Teacher Education) accreditation, our college had developed signature assignments to track students' growth over the course of their teacher education program. However, the emphasis slowly shifted from using signature assignments for informational purposes to ensuring student outcomes.

As we shifted from NCATE to CAEP (Council for the Accreditation of Education Preparation), we started having conversations about curriculum mapping, and standardizing student outcomes and even course syllabi. Our conversations in faculty meetings increasingly centered on data and became increasingly less democratic.

Some have warned of the attack on academic freedom, which if properly protected would support research philosophies like constructivism and child-centered practice. Henry Reichman (2015) states that academic freedom in higher education is highly vulnerable. Another scholar suggests, "Universities should be champions of academic freedom, and thus would be expected to support their faculty [who] must be protected from both external and internal pressures to allow for a more open flow of ideas and facilitate a diversity of opinions (Legates 2016, p. 23).

During my early years at the university, I experienced the protection and support for differing research paradigms and teaching philosophies. As time progressed, I at one point became the only full-time elementary science education faculty member in my college. During this time, I was asked to create a master syllabus for the elementary science course for others (mostly part-time instructors). I received emails from other faculty asking me what textbook I used for my course, and was happy to tell them I did not use one specific textbook, but relied upon multiple electronic and print resources including peer-reviewed journal articles. However, the part-time faculty informed me that they had been instructed by administration to use the exact course materials I used, and it had to be a textbook. Despite being the authority, I saw the degradation of university faculty and the limitations of academic freedom. When a few other faculty began to advocate for more shared, democratic governance and were questioning the direction in which our accreditation directives were taking us, one administrator said, "Faculty governance means faculty stepping up and taking responsibility for enacting the directives of our accreditation." As Common Core State Standards were adopted, our college put together a "task force" to make sure we were fully preparing our students for teaching these standards.

Despite my resistance and the resistance of others, students were seeing their teacher preparation program transform into a technical training institute. Pecorino (2013) suggests, "There are movements to transform higher education that would have its aims reduced to the nearly singular focus of short term job training and the production of employed graduates, quantifiable and publically transparent outcomes" (para. 2). Some of my students were beginning to ask about the feasibility of teaching divergent, child-centered, inquiry practices, commenting, "We really like the idea of authentic inquiry, but in reality, we will never be able to use this in schools." For the most part, they were right. Slowly, the academic freedom on which higher education should stand was eroding, and students who should be learning the art of teaching and strategies that meet the needs of learners were learning how to conform. One student said it best when interviewed. She mentioned there is no room to be a different kind of teacher in the teacher education program, and that "My grade is affected by my ability to conform." The controlling practices of K-12 practice had entered higher education. There is a hierarchical expectation to accomplish the goals of standardization, and each level has its own set of accountability measures to foster compliance and silence dissent (Canestrari and Marlowe 2005).

15.10 Pathways for Discourse and Challenging the Status Quo in Science Education

A colleague and fellow university faculty member recently spoke in a meeting on the issue of CAEP accreditation directives that directly impact the curriculum, instruction, and assessment in our teacher preparation program. She said, "We cannot frame faculty participation as either resistance or compliance because I happen to agree with the structures and direction [of CAEP], and I do not want to be labeled as 'compliant'. I do not think anyone should resist." In pondering this statement, I think of all those teachers and university faculty operating under an oppressive, high-pressure, performance-based system. Many agree with the mandates of the system, to which they have every right, but what about the voices of those who are opposed. There are many who disagree, and are inclined to resist, like myself. So, I pose the question, is there room for dissent? Is there room for teachers to be different? Can a science teacher step aside from standards-based, convergent, superficial content memorization tasks long enough to build students' authentic inquiry, processes skills, and a unique identity as a scientist? In my experience, the answer to these questions has most often been a resounding "no."

However, I do have hope and believe there are pathways for building meaningful discourse while challenging the status quo. My students, who have powerful voices to affect change at a variety of levels, give me hope. Lutzker (2007) describes teachers in the following way:

> Through realizing possibilities within a framework in which qualities such as *openness, sensitivity, flexibility, creativity and expressiveness* are deemed as most essential, the teacher as an artist in her classroom is seen as exhibiting comparable forms of skill and grace as a musician, dancer, or actor. This is a perspective with potentially far-ranging consequences affecting all aspects of teacher education and teaching. (para. 5)

I believe the conversations around academic freedom, classroom teaching, and treating science education as more than "cookbook science" (Demir et al. 2010), need to continue in the university classroom, in professional development opportunities, and in scholarship. I believe that teachers, professors, parents, and administrators need to form stronger networks to collectively push back against or terraform a hostile system. I have struggled to work against such a system, and have often felt alone in my efforts. What I have found is that one cannot terraform an entire system alone, and especially in isolation. Also, multiple people cannot terraform a system without communication and a strong network. It is easy for the system to squash agents operating in isolation. The teacher needs to be treated as the expert and the artist, and needs a strong, connected network of support. If we can strengthen a collective voice of dissent, and we find just small ways in our local systems to challenge the status quo, then we may have a chance at terraforming the education system.

References

Bloom, J. (2006). *Creating a classroom community of young scientists*. New York: Routledge.

Bochner, A. P. (1994). Perspectives on inquiry II: Theories and stories. In M. L. Knapp & G. R. Miller (Eds.), *Handbook of interpersonal communication* (pp. 21–41). Thousand Oaks: Sage.

Canestrari, A., & Marlowe, B. (2005). From silence to dissent: Fostering critical voice in teachers. *Encounters, 18*(4), 41–46.

Coughlin, P. A., Hansen, K. A., Heller, D., Kaufman, R. E., Stolberg, J. R., & Walsh, K. B. (2000). *Creating child-centered classrooms*. Washington, DC: Children's Resources International.

Darling-Hammond, L. (2011). *Linda Darling-Hammond's fiery commencement address*. Retrieved from http://blog.nationalequityproject.org/2011/05/25/linda-darling-hammonds-fiery-commencement-address/

Demir, A., Schmidt, F., & Abell, S. K. (2010). Science from the pond up: Using measurement to introduce inquiry. *Journal of College Science Teaching, 39*, 23–27.

Du Quesnay, H. (2003). Teachers and leaders – NCSL's part in developing the teaching profession. *Education Review, 16*(2), 49–54.

Einstein, A. (n.d.). *Education quotes*. Retrieved from http://www.froebelweb.org/web7011.html

Ellis, C., Adams, T. E., & Bochner, A. P. (2011). Autoethnography: An overview. *Forum: Qualitative Social Research*, 12(1).

Essex, M. (1952). What does academic freedom mean for elementary and secondary teachers? *Educational Leadership, 9*(4), 237–242.

French, D. P. (2005). Is academic freedom a threat to teaching introductory science? *Journal of College Science Teaching*. Retrieved from http://www.nsta.org/publications/news/story.aspx?id=51143

Haigh, M. (1998). *Investigative practical work in year 12 biology programmes*. Unpublished Ph.D. thesis. University of Waikato, Hamilton.

Kohn, A. (1993). Choices for children: Why and how to let students decide. *Phi Delta Kappan*. Retrieved from http://www.alfiekohn.org/article/choices-children/

Kuhn, D. (2007). Is direct instruction an answer to the right question? *Educational Psychologist, 42*, 109–113.

Legates, D. R. (2016). The university vs. academic freedom. *National Association of Scholars*. Retrieved from https://www.nas.org/articles/the_university_vs._academic_freedom

Lutzker, P. (2007). Teaching as an art. *Humanising Language Teaching*, 1. Retrieved from http://www.hltmag.co.uk/jan07/mart05.htm

Michael, T. (2012). *In diversity there is beauty and there is strength*. Retrieved from http://someonetotellitto.org/in-diversity-there-is-beauty-and-there-is-strength/

Moyer, R. H., Hackett, J. K., & Everett, S. A. (2007). *Teaching science as investigations: Modeling inquiry through learning cycle lessons*. Upper Saddle River: Pearson Education.

National Aeronautics and Space Administration. (n.d.). *Terraforming Mars*. Retrieved from http://quest.nasa.gov/mars/background/terra.html

National Science Teachers Association. (2010). *NSTA position statement: Principles of professionalism for science educators*. Retrieved from http://www.nsta.org/about/positions/professionalism.aspx.

Pecorino, P.A. (2013). *Academic freedom and CUNY: Insulty to injury via pathways*. Proceedings from AAUP conference. Queensborough Community College, CUNY. Retrieved from http://www.qcc.cuny.edu/socialsciences/ppecorino/PRESENTATIONS/AAUP-Conf-2013-Pathways-AF.html

Peters, J. M., & Stout, D. L. (2006). *Science in elementary education: Methods, concepts, and inquiries* (10th ed.). Upper Saddle River: Pearson Education.

Pinnegar, S., & Erickson, L. (2010). Teacher-centered curriculum. In C. Kridel (Ed.), *Encyclopedia of curriculum studies*. Thousand Oaks: Sage Publications.

Ravitch, D. (2010). *The death and life of the great American school system: How testing and choice are undermining education*. New York: Basic Books.

Reichman, H. (2015). Does academic freedom have a future? *Academe, 101*(6), 1–6.

Robinson, K. (2013). *How to escape education's death valley*. Retrieved from http://www.ted.com/talks/ken_robinson_how_to_escape_education_s_death_valley.

Schreiber, L. M., & Valle, B. E. (2013). Social constructivist teaching strategies in the small group classroom. *Small Group Research, 44*, 395–411.

Stone, B. (2014). *Influence of teacher-directed scientific inquiry on students' primal inquiries in two science classrooms*. Doctoral dissertation, ProQuest Dissertations and Theses. (AAT 3682808).

Taylor, P. (1996). Mythmaking and mythbreaking in the mathematics classroom. *Educational Studies in Mathematics, 31*, 151–173.

Vinson, K. D., Gibson, R., & Ross, E.W. (2001). High-stakes testing and standardization: The threat to authenticity. *John Dewey Project on Progressive Education, 3*(2).

Wallace, T., & Brand, B. R. (2012). Using critical race theory to analyze science teachers culturally responsive practices. *Cultural Studies of Science Education, 7*, 341–374.

Zion, M., & Mendelovici, R. (2012). Moving from structured to open inquiry: Challenges and limits. *Science Education International, 23*, 383–399.

Dr. Brian Stone is a Senior Lecturer in Teaching and Learning at Northern Arizona University. His research interests include multi-age education, child-centered practice, constructivism, childhood development, authentic assessment, inquiry-based science methodology, academic freedom and democracy in education, and critical pedagogy. He teaches undergraduate courses in science, social studies, math, and literacy methods, as well as evaluation of learning. He also teaches integrated curriculum and issues in multiage education at the graduate level. He was an intermediate multiage teacher (4th, 5th, 6th grades) prior to working in higher education.

Part II
Science Education, Politics and Resistance

Chapter 16
Not "Real" Science Education Research: The Systematic Silencing of Critical Science Education Scholarship

Jean Rockford Aguilar-Valdez

> What you write about is just little stories. Why should anyone care what those illegals think? How are they even in school? We're science ed scholars, not soap opera writers. – Researcher who expresses interest in equity and science education

> I don't know why those science teachers tolerate all your talk about racism and imperialism. They don't have time for that, they're busy teaching science. It's a wonder they don't rise up against you. – Veteran professor of science education

My counterstory tells of my experience as a Latina doctoral student and assistant professor in science education, in Teacher Education departments and Science Education conferences that are dominated by white and middle/upper class academics. As a Latina from a working class background, I often find myself out of sync with and ostracized by the culture of academia in general and science education academia specifically. I have had to learn how to survive the systematic racist, sexist, classist, ethnocentric, and elitist structures inherent in these worlds, even as members of these structures label me and my scholarship as unscholarly and try to domesticate me to fit the standard of what "real science education research" looks like. This meant erasing and belittling what I am, what I believe, and what I dedicate my life's work toward.

My scholarly work critically examines the structure of science education through highlighting the voices of those traditionally marginalized, using Latino Critical Race Theory (LatCrit) and Chicana Feminist frameworks. It often underscores the colonizing nature of science education as it is often practiced, and advocates for anti-oppressive, decolonizing methods. My resistance to mainstream narratives of science education research (which paint this field as colorblind, objective, and meritocratic) not only ostracized me from the "in crowd" within dominant structures of

J. R. Aguilar-Valdez (✉)
Portland State University, Portland, OR, USA
e-mail: Aguil@pdx.edu

J. Bazzul, C. Siry (eds.), *Critical Voices in Science Education Research*, Cultural Studies of Science Education 17,
https://doi.org/10.1007/978-3-319-99990-6_16

science education, but justified their continual abuses and oppressions of my voice, by body, and my research. My counterstory is intended to expose the racist, sexist, ethnocentric abuses of power underlying academia within science education, often shrouded under a patriarchal and "white savior" lens of tokenism at best, and forced assimilation and gatekeeping at worst.

The underrepresentation of women and people of color in science, science education, and academia seems to be a strong focus of those claiming to work for equity in these fields. And yet, the structures of white supremacy, patriarchy, classism, and elitism within science education research remain in place, and the holders of power in these fields continue to make women and people of color feel unwelcome in these worlds. This is especially traumatic for women of color scholars like myself when the very people making the field hostile and unwelcoming are those that are defining their scholarly careers around issues of equity and social justice in science education. Further, many of those science education scholars who do not outright reject this kind of scholarship, take it upon themselves to paternalistically "counsel" me to soften my words and censor my findings in order to appeal to the tender sensibilities of those that dominate the field, and not "rock the boat." This attempt at silencing is seen through their eyes as helping the poor, misguided brown scholar to assimilate to the world of science education, so that she can "fit in" and "make it" – an act of domestication shrouded in white saviorist "good intentions."

Through my narrative, I will share my experiences as a doctoral student and later an assistant professor, and one of very few Latinas entering the field with years' experience as a scientist (in physics) and science teacher to groups of low-SES Latin@ middle school students. I brought into my research a passion for teaching science in ways informed by my own working class upbringing and ethnic identity. The message I continually receive from many scholars in science education is that my research and identity is offensive and contrary to "real" science education and "real" education scholarship, which they claim to be objective and without the influence of white supremacy, patriarchy, ethnocentrism and classism. Efforts to teach me "my place" included telling me I wasn't a "real" science education scholar because of the lenses I invoke and the critiques I use, and that my work is "dangerous" and I should "be careful whose toes [I] step on" and what bridges I burn.

The old guard systems of power are alive and well within science education research, and bent on maintaining the status quo. In telling my story, I aim to shine a light on the systems that propagate the oppressive underbelly of science education research and hope to actively resist it. I hope that in telling my counterstory, others from historically marginalized backgrounds who have been oppressed and abused by the status quo within the academic world of science education, will feel solidarity, will empathize with shared patterns of oppression, and will feel compelled to speak out in chorus with me against the traumas we've endured. I also hope that those involved in science education research will take note of these experiences in order to raise awareness that this climate exists, and work together with the marginalized, to change it.

In addition to my story told through narrative, I will illustrate the slow and steady systemic "domestication" of this Latina scientist and science educations scholar through a "death by a thousand cuts" illustrated well by Frida Kahlo's *Unos Cuantos*

Piquetitos (Apasionadamente Enamorado) [A Few Little Cuts (Passionately in Love)], that works to render us invisible while bleeding us dry, the image is shown below:

To set the scene: it's important to realize that my work lives on the bridges between science education and Chican@ feminism, where I work with Latina students in middle and high schools, ethnographically learning about their complex relationships with science education, which are overarched by their own complex identities and the ways they must navigate political, social, and educational structures fraught with power dynamics and the effects of colonization. I also work with teachers of all subjects, particularly science teachers, on an ongoing basis to disrupt colonizing approaches to science education, and open up spaces for socially just praxis that broadens the myopic Western definitions of what counts as science and what counts as teaching and learning. I have Bachelors and Masters degrees in physics and chemistry, and prior to receiving my Ph.D. in science education and becoming an assistant professor, I was a middle school science teacher in East L.A., in a Title 1 school that served predominantly Latin@ students. In my current scholarship, I also authoethnographically reflect on my own struggles as a Latina from a working class immigrant upbringing who is now in academia, and the ways that my identity, research, and outlook interact with an academic system still very much informed by systems of white supremacy, patriarchy, and other colonial mindsets. My, and my students', women of color ways of knowing are in constant contention with the systems that many of my students and I navigate. I consider my work to be in solidarity with that of the Latin@ students and teachers I work with, as they have just as much, if not more, to teach me that I could ever offer them. The following reflects my struggles and journey, as I reflect on my past and my present worlds. I will leave you now to let these words sink in...

Unos Cuantos Piquetitos – By Jean Aguilar-Valdez

I hoped for escape from all that I was
Sold on the stories they told me
That education was my salvation
And that I needed saving.

My culture, my language, was una Vergüenza.
Uneducated
Backward
Shameful

And so to prove myself, I needed more
Shakespeare, Hemingway, Newtonian Mechanics, Euclidian Geometry
In my pathetic Marinera life
To erase the stench of the Cuban Balsera
To erase the stench of pastelitos and Fabuloso floor cleaner
Hope, they said, was college, education
Science and Math, the language of success
Defined on their terms
Using the old white male rules
I memorized their names
Their theories
Their equations
Their periodic tables

Because
I hoped to fit into their community.
Because
My parents came to this country so I could have a future.

Because
Your future is carefully measured by how many
Questions you get right on the standardized tests
and
Scholarships you get to ivy-league universities

And so I studied them carefully
The "successful"
I learned to walk like them
Talk like them
Think like them
Know like them

I hoped to fit into their community.

By cutting away all I was
And becoming all I hoped to be:
One of them.
The funny thing about cutting
Is that it leaves marks that bleed
And scar over
To remind you of the cut.

And no one ever warned me
When they pointed out this golden path to salvation
From all the terrible things that they told me I was
No one ever quoted the *Katha Upanishad:*
"The path to salvation is sharp as a razor's edge, impassable, and hard to go by"
They just said to go, make something of yourself
And the definitions of "something"
Were defined by the white men
In the white coats
With their learned gazes

And my little Cubanita heart ached for their approval.

I hoped to fit into their community.

What happens with the tender hopes of Latina flesh
Pass over the razor's edge?
Into their community?

El Primero Piquetito:

I'd never seen a science lab before
The smell of chemicals and old metal and glass equipment
The old cloth-bound books on the shelves
With titles like "Advanced Thermonuclear Physics" and "Non-Newtonian Fluid Dynamics"
Excited me
It was nothing like the novelas and platanitos of home

I made it.
I am one of them.
My dream realized
A *scientist.*

And I asked a professor what I had to do
To declare my major in physics
And he said:
"Physics is really hard.
You sure you don't want to major in English instead?
Girls aren't well suited to the rigid logic of science
And besides
Hispanics could always use
A little more English."

<u>El Secundo Piquetito:</u>

The only female in my quantum physics class,
I raised my hand often to give answer after answer
Having studied so hard
To prove that I belong here
So I could be accepted as part of their community

And the male students whispered behind me
"How does she know all this?
I bet she's sucking the professor's dick to get all the right answers."

<u>El Tercero Piquetito:</u>

Working in a science lab
It was all white males and one other white female, and me.
Frumpy, nerdy invisible Latina me.
And the males started ranking the fuckability of the white woman in the lab:
"Would you do her?"
"Yeah, she's a hot little thing."
"I'd do her from behind while she wore her cute little lab coat."
I learned to be silent.
Women of color in these situations learn to survive by being silent.

And a white male science educator
Quite renowned in his field
Told me that I need to stop my whining
"This is why people like you don't belong in science
You're too emotional.
Get over it.
This multicultural stuff you talk about, it belongs in a philosophy department.
It's not science and you're not a scientist.
I'm surprised the students don't rise up against you
You need to stick to teaching the subject you were hired to teach."

Pero sólo fueron unos cuantos piquetitos

And this hopeful little Cubanita from Miami
Pounds her raw fists along the razor-wired edges of their world
Leaving blood stains outside the closed gates of their community.

And finds a new one, filled with women of color Latina Chingonas
with scars as numerous as hers,
with voices as quivering as they are loud
louder now in solidarity
mujeres cansadas from loss of blood pero
no terminadas
not by a longshot

And here
She can heal.

Jean Rockford Aguilar-Valdez As Latinx, a scientist, and a science educator, I am personally familiar with the inequity that exists for people of color and those that are not cisgendered males in the science and science education worlds. I'm a community-engaged scholar-activist that works at the intersection of Chicanx feminist theories and science education, to push against colonizing, narrow conceptions of science and science education, and advocate for socially just, decolonizing understandings and praxis regarding the teaching and learning of science.

Chapter 17
In the Middle of Treaty Walking: Entangling Truth, Ethics, and the Risky Narratives of Two Settler(colonial)s

Audrey Aamodt and Jesse Bazzul

17.1 The Middle of ......

Audrey: Hey Jesse, let's just be in the middle. I detest having to start somewhere and finish somewhere, always coherent. Who cares about coherence? Well, maybe a lot of people do. Jesse, just say something, start anywhere. Jesse, are you awake?

Jesse: [Day-Dreaming] *Follow the Map Audrey, the Map! Like Coyote and Raven* (O'Riley and Cole 2009)*!* What? Oh-uh. Sorry, I think I was dreaming. You were placing different things: buildings, stalks, cars, and little animals, on a wide-open field. Well, ah, maybe even knowing about coherence is enough for it to work on us. I feel like we need to walk out of institutional coherence, out of this small world. Which is why I was asleep maybe? So what did I say?

Audrey: You said, 'follow the map'. Remember, we've been talking about mapping. Yet, what if mapping doesn't necessarily tell us where to go; something less straightforward than a google map route from here to there. Sure we can make traces.... and, can we tangle it a bit? Becoming-ethical as a tangle; I'm thinking about *mis-steps* on our treaty walking.

Jesse: Well... if you're sure this isn't a dream…

In order to make sense of *mis-steps* I think it might be useful to tell about where we've been, to try to show people. It might help us feel contours, and play with their importance. We might also wonder if *mis-steps* exist; maybe all steps are intentioned by places, things, and forces. You may not remember, but we got here (to this spot, to this text) because we/they... I mean, what do you dream about?

Audrey: Tangles. Strawberry roots, all a tangle! Remember those little metaphor-stories we told at the Treaty Ed Gathering? *So, since we've last met, my treaty walk has been like strawberry plants. The roots grow this way and that, with beginnings*

A. Aamodt (✉) · J. Bazzul
University of Regina, Regina, SK, Canada
e-mail: Audrey.Aamodt@uregina.ca; Jesse.Bazzul@uregina.ca

J. Bazzul, C. Siry (eds.), *Critical Voices in Science Education Research*, Cultural Studies of Science Education 17,
https://doi.org/10.1007/978-3-319-99990-6_17

always in the middle of things. Some shoots could be clipped and placed elsewhere or given away to someone else; In that patch of soil over there; that place that seems like a bare spot. But what if we let some others go wild, before we gnaw them off with the weed whipper! And, then again, maybe there are no bare spots and never ever have been. That was my mother's way of talking in the garden.

Argh! To hell with making sense! To show people What, exactly? And Why, exactly? Well, the why part... I suppose we have a why in mind, with a wink and a nod. For, as you said, *I want my life to be ethical... It matters to me more than anything. What could matter more?*

Jesse: Ok, ok. Let's imagine. Let's show people a treaty walk, this Treaty 4 Land. This all started from that time in the science lab when we posed the question of an ethical self. I think you're ahead of me, in time, space, and even this text, but... [*whispering*] nobody can tell, so it's ok... and the reader may soon overtake us. We both want to come to a space where we do a 'good' thing. We seek a relational way; a different way to think about ethical things; about doing relational ethics differently. Because it comes down to realizing we are *poorly understood selves…* [*Audrey hmmmms; Jesse laughs*]

…which explains our creepy attachment to Foucault's (re)conception of Parrhesia – a dead concept (from dead white men)? A zombie concept revitalized as truth telling for the purposes of what is correct/good and self-examination. Let's think about this. We were calling it 'the Land of Parrhesia', in all our colonial ridiculousness, truth-telling on a map, and with all the terror, risk, and wonder truth allows. Like, in Canada, Truth, in part, must remain on a map (of virtual of possibilities, but also legacies of European colonization) – nothing can proceed without a reckoning with Treaty Maps.

Truth-telling for the self is happening here and now as we're on our treaty walk; where we think about self, truth, reconciliation, decolonization, and what we will do/are doing as settlers on these Indigenous lands (treaty walk people). This is a Parrhesiastic problem of historic proportions – for it is Canada's 150th birthday this year. We shouldn't be looking forward to the party.

Audrey: Yeah, let's crash that party. Disrupt the festivities.

What if we asked those Canadian party animals, those people like us (or not): *Since we last met, what has your treaty walk been like?* I wonder what they'd say. Well... what do you say, Jesse? [*glancing over my shoulder*]. I notice you're beside me... maybe being ahead on the map, in space and time and text, on this road to/of becoming-ethical is a myth. Ahead, getting left behind, overtaking one another... these all remind me of the binary of good/bad.

Jesse: It's ok, we are always reorienting always, and treaty walks can be out of phase. From the middle we can draw things together and create an assembly of things. An ethical-assemblage-journey and a line that runs right through it. We sketch from time to time, like kids might on bark strips (Bazzul and Santavicca 2017), and return it to the map. We need to put ourselves into the picture because the "Courage of Truth" is in the telling … the ethical examination of self and other.

Audrey: "Like kids might on bark strip"? I wonder why bark strip? What are you referring to here? It seems like an attempt to include Indigenous Knowledge or traditions – is this a token? A stereotype? Another mis-step?

Want to make a pact? Like a parrhesia pact? Will we risk death of the self; this current self? The death of Canada, as we think we know it: "to risk the bond between the person speaking and the person to whom he speaks" (Foucault et al. 2012, p. 13)? This process of constituting ourselves as truth-tellers, as speaking freely to one another takes courage. What could it look like to make such a pact? A pinky swear? A handshake? Written in blood? Blood brother-sisters. Beyond flattery and gentle, honeyed comfort. Then, blunt? Harsh? I'm thinking of conversations with my students about considering what settler-invader (Newbery 2012) might mean to us/them, and the ways they think this is harsh language, squirming in their seats with tight lips. Is it just a game? Just a game of truth questions?

Jesse: Risk Death? Yes. [*But is this a dream or not? Things cannot die in dreams...*] You cannot interrogate through truth-telling (Parrhesia) without risk to one's self – which includes the risk of truth turning back toward the self as you say. What desire do we have to enter this game and become ethical? I think our pact may not be with each other, but between us and ourselves. Do we risk the self being unraveled? The fact that we are here in the middle suggests we already have. The drawing on bark came from my queer trip to Camp Wilde (Gough et al. 2003) – it was a childhood practice, something perhaps needing psycho-spiritual-political examining (Bazzul and Santavicca 2017). But, take me on a treaty walk, I would like to stray backward now.

Audrey: I hereby promise to risk the self being unraveled!

This dream-story, *Homesteading Home*, is one of my games of truth... I don't know how to map stories like these.

> *I'm at Grandma Aamodt's farmhouse. A large silver bowl rests on the side counter. Actually, it's not silver – I just remembered – it's one of those blue flecked porcelain coated bowls with handles on each side, and a bit of a lip with a sort of built-in stand: a settler bowl for making bread dough. Quietly, I'm watching. She's pouring canola oil into the bowl, coating the sides so that the dough won't stick while it rises. Her pale hands, fingers gnarled with arthritis at the joints of her forefingers, seem to know how the dough should feel when it's ready. With flour and water and egg, her hands knead the ball together, as I stand and watch and wonder about how she knows, without exact measurements or a written recipe, how to make bread. How many times has she done this before? I'm her oldest girl grandchild – maybe I'll also make good, wholesome bread when I'm grown up.*
>
> *It feels special to watch her. It feels not so long ago that we'd have to drive a torturous, 6 hours across Saskatchewan in order to visit her. Suddenly I recall the waxy scent of left-behind melted crayons, the carsick inducing smell of the backseat of the maroon '86 Oldsmobile. We used to travel here yearly for Christmas, or for a summer visit. Now, living only a mile and a half from her, I get to be here whenever I want. I've even been brave enough, once or twice, to walk across the fields that are between our yards – when there isn't crop growing there. My dad grew up on this farm, eating my grandma's bread and helping his dad do chores and work in the fields. He always wanted to be a farmer, and so that's what he is. He was farming near Reward with my other grandpa before we moved here. That was where my mom grew up, where my great-grandpa Zunti homesteaded. But maybe it's important for him to be living once again at his childhood home; To be farming the land that his dad and grandpa also farmed, closer to his own family's homestead.*

Sometimes I'll ride beside him in the combine, while it picks up the swaths of yellow-golden wheat. The dust is rich with some kind of scent that I can't really describe. Maybe it just smells like wheat field in the air. Through the little, dusty back window, I watch the kernels flow into the hopper like water. My dad's head swivels forward and back, constantly checking to make sure everything is okay. The noise from the front, where the swaths are being chopped inside the belly of the combine, is muffled by the cab which protects us. Now and then, dad pushes a lever this way or that and when the back is full, we wait for my uncle to bring the grain truck beside us before we can unload. It's a blue-cabbed one, with the red wooden box; paint faded and chipped; My grandpa's old truck from the farm where my mom grew up.

When we're outside, it's a habit of Dad's to give me some wheat to chew – to make wheat gum. He does it too. The little handful of seeds is crunchy at first; not sweet but earthy, as it gradually turns into a mush and starts to stick together in my mouth. It's not hit-me sweet like Hubba Bubba(TM). You can't make bubbles with wheat gum either. And it's not pink of course. Chewing and thinking: Thinking of the names we have for all the quarters of land we own and farm: we live on the "home quarter", the 160 acres that our yard and house is on, surrounded by waving fields of green in the summer and, now, yellow days of grain dust in the air and the roar of machinery as the wheat is gathered off the land; off the "Homestead"; off the "Walker quarter"; and others, into bins; to later sell or haul away.

The Walker quarter is our name we have for that piece of land where the Walkers used to live. I never knew them; never viewed a picture of these people called Walker. Their house is no longer there, just a hole that used to be the cellar. It's one of our makeshift landfills at our Farm. Another is near the entrance to the homestead, close to the treed edges, where there is a pile of old, broken TVs, mattresses, a freezer; a dumping place for broken things. I remember helping my Dad haul things to these spaces, filling the Walker house hole up with old shingles from our yard; or when an animal died, cow or sheep, bringing it there... calling it the dead animal pit. My mom and I would walk a little faster past that place on those warmer days when the smell would waft towards us on the road.

Landfills and landmarks. But perhaps, being a farmer's daughter is more about learning to make bread than to think about these darker things along with the concerns of combining. I keep wondering, will I just know how someday, like my Grandma does?

There's lots to talk about within this little story. I feel compelled to note the ways in which we name places at my family's farm. We name them our names, to say that they are ours; to shore up our sense of belonging. Well, and also because *Home*steads do that, they make it our home, not their or your home anymore. Especially when other stories have been ignored and it seems we have chosen to keep it that way. My family doesn't know any other stories about this homestead land. Instead, the farm-sReserve border is often called upon. This little memory peers into a moment before I knew about treaties; about contested land; about truth, reconciliation, and being a treaty person on a treaty walk. When I was 12, First Nations People lived in my mind as long ago, nomads living in teepees with buffalo hunts and hides. And that's where they stayed, in the past, in my mind. Here I was and we were, *homesteading home* with bread and wheat and family, very near to *Muskowekwan First Nation Reserve*... only about 20 km on grid roads... without this attention. It's bizarre. Unethical. But useful for becomings-ethical, becoming-treatylandperson.

In such a context, I've struggled with the concept of "ethical relationality": "an ecological understanding of human relationality that does not deny difference, but rather seeks to understand more deeply how our different histories and experiences position us in relation to each Other" (Donald 2012, p. 535).

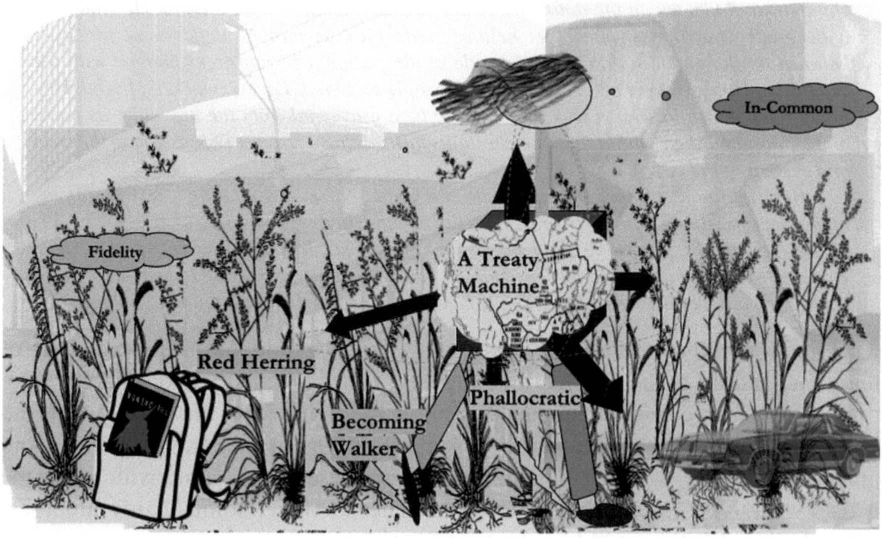

Fig. 17.1 Body-privileged-male-white-on-a-treaty-walk

I tell stories like these as if we belong here, like they are normal, and we are normal to be here. Simultaneously, they are and they aren't; we do and we don't. The unsettling truth-telling about this story is that this land wasn't, and isn't, ours.

Jesse, take me along on a road of your treaty walk.

Jesse: I am tempted to talk of the treaty as a machine; an organizing, producing machine – for settler-colonizers (Veracini 2011) treaty machines have served to both destroy and preserve what we share in-common. Me-on-a-treaty-walk is also a machine operating in a white phallocracy. So OK, yes, let's start with this kind of machine, a body-privileged-cismale-white on a treaty walk. Its organs are wasting, with new beginnings; the treaty machine is deaf and comic. The machine might look like this (Fig. 17.1).

17.2 Dreams of Nauky: Fidelity and Being Communal

Out of time, or somewhere in future-time, I am sitting under a tree. Someone has guided me there, and softly tells me to sit and begin writing. My first sentence: "Treaty is a preference for community over conflict. To embrace treaties is to embrace fidelity to a future". But I realize my treaty walk also involves infidelity, breaking with what came before. Thinking of both, there are times when the world opens up and something big happens.

Some years back, I lived in Kyiv, Ukraine. What struck me most, at first, was the uniform surface of things – a uniformity built by a great state power. However, after some time, I began to notice the disorderliness that lay underneath this uniformity. Behind my Soviet-style block apartment on Prospect Nauky (science street) was an old climatology station that was slowly being taken over by chestnut trees. On the backside of the red-bricked building (just so you know, red bricks do not exist in Kyiv), was a brightly painted green door. Everyday, children would knock on the door, and sometimes an old man would answer. He could not understand what they wanted; yet neither could they, seemingly, understand why they knocked on the green door. On the far side of the station, chestnut trees merged with the expanse of Holosiivskyi Park. A place where people gathered. Lovers, workers, friends: the hungry people. Holosiivskyi was rumoured to be a graveyard from the big wars, its paths and tombs unmarked. In Autumn, the chestnuts fall, adding a strange percussion to the steel and cement of the city.

–Here and now, a piece of paper with the words "a second reading", sits beside me under the tree; along with the words, "fidelity or infidelity".–

As the months passed in Kyiv new signs moved in: Toyota™, NATO, & American Express™ It added a different order to things. And, eventually, we began to feel out of place again. The quiet of the weather station was fading. Chaotic dreams coming to a close.

Going forward as someone-who-dreams, I feel it important to ask something of that time, and all times that feel like dreams. Times of fidelity and infidelity. The time hidden from capital. What do they allow me to hear and see? Why is time so vivid sometimes? On a treaty walk my mind returns to the site of dreams. And maybe this is why we need a map, because all these things come with us on the walk. I want to know where my future dreams lie; as Alain Badiou (2014) might say, where love and politics reside.

A machine begins to move, its organs fading. It slides upon the path heavy, uneven, and pale. It pokes its head through Walker's corner, at the bottom of the landfill.

Audrey: *The machine of the body (cis-woman, white, bloody) begins to move, its organs failing.* These organs burst during settler mis-steps – naming myself as white-settler, treaty people; Knocking on the bright-green door; an opening of generational unknowns and unknowables.

I feel a dilemma: How can we challenge white-settler innocence, reconstituted by white spaces like *Canadian wilderness* and *homesteads* and *capital*, with more than just lip-service? Perhaps these treaty walk stories are useful in what we do with them. The expanded vividness of past-present-future living moments.

Jesse: Dreams of a multitude, replacing the old dreams with new ones. Chaotic dreams can turn to nightmares of enclosure – the biocolonization of knowledge, (the appropriation of Indigenous knowledges for the profit of a few), and the exclusion of knowledges other than Modern Western Science (Cajete 2000). Enclosures of the self in academia – being beholden to conservative fields like science education. To story this? It goes back to communal life for me. The absence of it. The rule of social and educational life by capital and hierarchy. It seeks to corrupt every moment we all share together.

Audrey: Disruptions of corruptions. Rhizomatic roots poke through the edges of nightmarish enclosures, searching beyond them; multitudes of gestures; efforts to consider the tensions in life.

Encouraged, I walked this morning, noticed an earthworm on the road, with mist in the air. Spending time outside, part of a collective gesture, calls me to challenge reassertions of white innocence & wilderness. This collective gesture is also about recognizing that I am a white-settler Treaty person living on Treaty 4 land, in relationship with more-than-human beings, along with, the *nêhiyawak* (the People of the Four Directions; perhaps commonly known to Canadians as the Plains Cree), the *anishinabe* (Saulteaux), the Dakota, Lakota and Nakota peoples as well as the four historically Michif (Métis) communities of Lebret, Fort Qu'Appelle, Willow Bunch, and Lestock, in this region, this place many call Southern Saskatchewan in the Treaty 4 territory.

Jesse: And when we go places with students, what and who will we recognize on the Treaty 4 land? These places have colonial names. It matters how we name or not name things. How do we keep our classes open to ethical and political possibilities at all times? For example, stand up for those students in Saskatoon who stood for Standing Rock (MacPherson 2016)? It is the pedagogical situation, in our science and environmental education classes, that will lead them from one space into another; from one ethical world to another. Education is the space of truth telling and ethical becoming, where people are brought to see speak about who they are, but more importantly who they will become in relation to creatures and things of the world. Science and Environmental education can create new ethical ways of being that move beyond human relations (though these are important).

> [The teacher's] truth-telling brings together and binds; the parrhesiast's truth-telling risks ***hostility, war, hatred, and death***. And if the parrhesiast's truth may unite and reconcile, when it is accepted and the other person agrees to the pact and play the game of parrhesia, this is only after it has opened up an essential, fundamental, and structurally necessary moment of the possibility of ***hatred and a rupture*** (Foucault et al. 2012, p. 25).

Audrey: So, in the interest of risking the truth and the process of risking it together, I also wish to take parrhesia risks with students. Teaching that risks the death of what we once thought, knew about ourselves, and relationships. I think of the hatred-hostility in students, when they are challenged to think of themselves as settler-invader, or white, or heteronormative, or able-bodied, or reproducing the gender binary, or to question this thing called objective knowledge and science education. In these moments, I risk the death of my science career. I think we must risk the death of who we thought we were, and what we thought science was. To me, this highlights the political process of teaching-as-parrhesia. For, "when scientific discourse is deployed as criticism of prejudices, of existing forms of knowledge, of dominant institutions, of current ways of doing things – and it cannot avoid doing this, in its very development – it plays this parrhesiastic role" (Foucault et al. 2012, p. 30).

Jesse: With science you also have truth playing out as a game, set in the contexts of other games, for example when truth is "told" through education (games). The search for a better way to be (ethical) may be inevitable, like death is inevitable, but also like how new life is inevitable. In this inevitability, we should recognize multiple forms of truth-telling and ethical modes of life. In my story from Ukraine, there is a truth of 'an event'; something born out of that time. In your homestead story, there is an evolving truth of the narrative atop the prairie. Maybe what an educator should do is to set in motion a series of truth modes – modes of parrhesia, and have these then lead to inevitable, yet unpredictable outcomes. Coming to terms with neocolonialism and white settler dominance involves something so radical – the return of very large tracts of land and the dismantling of current governance structures – that we must set in motion truth (machines) in our practice. We are braiding these machines while the (green) doors are rattling, when things are getting louder (Fig. 17.2).

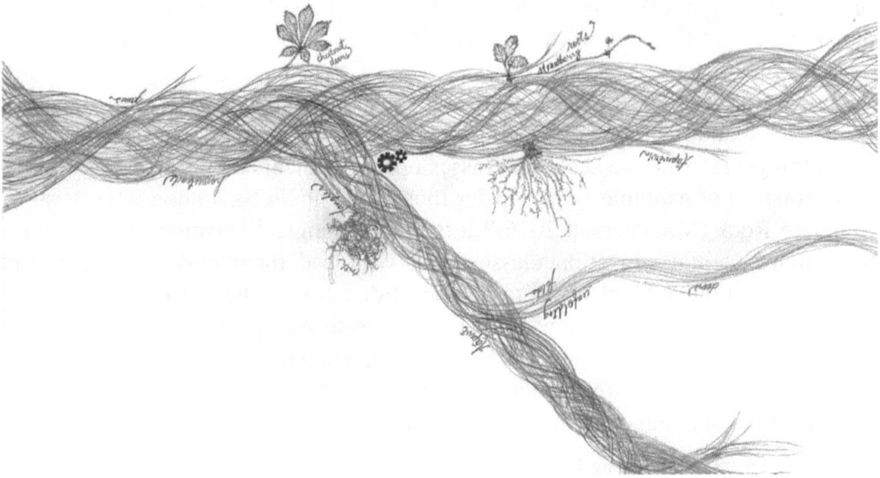

Fig. 17.2 Braiding truth-machines

Audrey: And, this treaty-settler-me can't just say, 'yep, I'm a treaty-person' and call that good enough. If I just use it to make myself feel better, to re-validate my claim to be here on this land, uncritically, without recognizing the numbered treaties as surrender treaties, as a taking away of land from Indigenous Peoples for settler-groups...then, this settler-colonization story continues to seek to assimilate; to make Canadians; to own the land. It's another move to innocence (Tuck and Yang 2012). So, instead, I notice the bright-green, rattling, unmarked doors; those potential thresholds of truth-telling, of *tâpwêwin* (translation: speaking the truth with precision and accuracy; Plains Cree language, *y* dialect, *nêhiyawêwin*); of treaty-settler identity that should always feel uncomfortable in this body-machine, uncomfortable in these truthgames... The doors! The doors! Open some, close others. A radical machine pops and slips and whirrs and pulls and grinds and (un) folds in *tâpwêwin* moments. Or perhaps this radical machine is something more subtle, or, then again, perhaps deafening. For, it's not simply reconciliation-talk to ease the national con**science**, but what will we **do** as we pass through these door-openings. This dismantling; It's screaming.

References

Badiou, A. (2014). *Infinite thought: Truth and the return to philosophy*. New York: Bloomsbury Publishing.

Bazzul, J., & Santavicca, N. (2017). Diagramming assemblages of sex/gender and sexuality as environmental education. *The Journal of Environmental Education, 48*(1), 56–66.

Cajete, G. (2000). *Native science: Natural laws of interdependence*. Santa Fe: Clear Light Pub.

Donald, D. (2012). *Forts, curriculum, and ethical relationality. Reconsidering Canadian curriculum studies: Provoking historical, present, and future perspectives* (pp. 39–46). New York: Palgrave Macmillan.
Foucault, M., Davidson, A. I., & Burchell, G. (2012). *The courage of truth: The government of self and others II; Lectures at the Collège de France, 1983–1984* (Vol. 8). New York: Macmillan.
Gough, N., Gough, A., Appelbaum, P., Appelbaum, S., Doll, M. A., & Sellers, W. (2003). Tales from camp wilde: Queer (y) ing environmental education research. *Canadian Journal of Environmental Education (CJEE), 8*(1), 44–66.
MacPherson, A. (2016). Student participation in anti-pipeline rally a 'significant concern': Greater Saskatoon Catholic Schools. Saskatoon Star Pheonix. Retrieved from: http://thestarphoenix.com/news/local-news/student-participation-in-anti-pipeline-rally-a-significant-concern-greater-saskatoon-catholic-schools.
Newbery, L. (2012). Canoe pedagogy and colonial history: Exploring contested space of outdoor environmental education. *Canadian Journal of Environmental Education (CJEE), 17*, 30–45.
O'Riley, P., & Cole, P. (2009). Coyote and raven talk about the land/scapes. In M. McKenzie, P. Hart, H. Bai, & B. Jickling (Eds.), *Fields of green: Restorying culture, environment, and education* (pp. 125–134). Creskill: Hampton Press.
Tuck, E., & Yang, K. W. (2012). Decolonization is not a metaphor. *Decolonization: Indigeneity, Education & Society, 1*(1), 1–40.
Veracini, L. (2011). Introducing: Settler colonial studies. *Settler colonial studies, 1*(1), 1–12.

Audrey Aamodt is a PhD Candidate at the Faculty of Education, University of Regina, Treaty 4, Saskatchewan, Canada. She is interested in how Deleuzoguattarian becomings work to unsettle the self through (un)braiding life-writing moments. This informs her curiosities about teacher education, settler-treaty education, and outdoor environmental education possibilities.

Jesse Bazzul is Associate Professor of Science and Environmental education at the University of Regina. He feels that comprehensive attention needs to be given to the way typically depoliticized fields of study, such as science education, work to constitute political and ethical forms of life.

Chapter 18
Engaging in Research Practices as Critical Scholars/Activists: A Metalogue

Alexandra Schindel, Sara Tolbert, and Alberto J. Rodriguez

Alexa: In taking a critical scholar/activist stance, we each have committed to social justice research to re/imagine possibilities in science education. We want youth and teachers to have significant opportunities to engage in learning and doing science in ways that (a) are productive, meaningful, and socially transformative, and that (b) provide youth with opportunities to succeed academically at all levels in science education. While many science education researchers work towards these goals, we turn the social justice lens inwards to be self reflective about research practices and consider the ways that these goals are negotiated and created within research relationships, structures, institutions, processes, and products.

We begin this conversation on a personal note to look at our positionalities – who we are, how we got here, and defining moments. What has it meant for you personally and professionally to engage in critical research as a scholar/activist? How does your identity influence your engagement as a scholar/activist? Have have there been experiences that have confronted or were affirming of this identity, and did certain people or spaces/structures/institutions help you to flourish?

Reflecting on this myself, as a White woman with a middle-class upbringing, who lived many years on the edge of poverty as an adult, I am acutely aware that my privileged upbringing has ensured cultural, racialized, economic, and social safety nets for me that are often unavailable to others because of the whiteness of my skin and the economic security of my extended family. I was moved to become a science teacher because of my love for science and because the social and racial inequities

A. Schindel (✉)
University at Buffalo, State University of New York, Buffalo, NY, USA
e-mail: aedimick@buffalo.edu

S. Tolbert
University of Arizona, Tucson, AZ, USA

A. J. Rodriguez
Purdue University, West Lafayette, IN, USA

J. Bazzul, C. Siry (eds.), *Critical Voices in Science Education Research*,
Cultural Studies of Science Education 17,
https://doi.org/10.1007/978-3-319-99990-6_18

I saw made me want to be a part of productive and redistributive educational change. After teaching middle school science for several years, I completed a PhD to research social justice within formal school science with teachers who work in primarily minoritized communities and with the youth in their classes. I examine the ways that teachers and youth experience empowerment and act as agents of change in this context. My lived experiences and identity drove me to a desire to conduct research embedded in school settings because the formal processes of schooling makes social justice endeavors both incredibly complex and potentially powerful in the lives of young people. However, after completing my dissertation and beginning work as a professor, I had not yet created ties to other like-minded science education researchers (such as Sara, Jesse, Alberto and many of the authors in this edited volume), and I felt in this isolation that my work was not sustainable. While my research was going well and I had established fulfilling personal connections with several teachers and youth through a research project, I felt isolated and like my work was obscure and marginalized within the field of science education. The major turning event for me was meeting several like-minded colleagues in science education who became my support system. Within these relationships, we affirmed our identities as scholars and activists, engaged in reflecting on our identities and positionalities in relation to our research, encouraged each other's research and activism, developed connections with others to support critical science education scholarship and community, and supported one another through significant stresses that occurred when colleagues questioned our activism and/or scholarship.

Sara: I can relate to Alexa's experiences of privilege as an Anglo European, middle class White woman. My relatives on my mom's side grew up in Appalachia in the U.S.. Many of our family members on her side were coal miners. My grandfather, in his youth, went to work briefly in the mines and hated it, so he joined the military as the most tangible path for pursuing social and economic mobility. My grandmother on my mom's side, who grew up in the Midwest, did not complete school. She stayed home to care for her 12 siblings, though later went on to pursue a GED and then became a licensed nurse practitioner. My dad's father came from a carpenter family and his mom was from a family of small farmers, mostly subsistence farming. She did go to college and majored in Home Economics (few options for women then). My grandfather went to seminary and became a minister and seminary professor. My grandfather was ultimately pushed out of the Southern Baptist Convention for his non-literalist views of biblical teachings and refusal to subscribe to the inerrancy doctrine. Ironically, despite his refusal to conform to certain societal and professional standards he viewed as unethical and unreasonable, my grandfather was always worried about people viewing his children as unintelligent because of their, as he saw it, "unsophisticated" roots and southern accents. He worked hard to train his children, including my father, out of speaking with a thick southern drawl and emphasized education as a means for social mobility. (My dad did the same for me.) Essentially, however, because we were White, we could fairly easily shed these markers and visible traces of our "low class" origins – a function of privilege inaccessible to Black or Latino families in similar situations.

My mom had gone to nursing school as one of the few "legitimate" ways to escape an abusive household – and because nursing was one of the few career options for women. She met my dad shortly thereafter, and they married early so that my mom would not have to go back and live at home after nursing school, as single women in the south were expected to do. My mom became a nurse, and my dad was a chemistry professor, and I, along with my three siblings, was raised to believe that I could do whatever it is that I set out to achieve – as long as whatever I set out to achieve was done with integrity, and with care for others. I see this as both (1) a form of anti-patriarchal resistance on the part of my parents, who raised very empowered and socially engaged girls (three of us) who all went on to get PhDs in science-related fields (chemistry -oldest; science education – me; comparative biomedical sciences – youngest), as well as (2) the insidious nature of White privilege in this country – in that this "American Dream" is not intended for everyone, by design.

I did not know what I wanted to do after high school (or during college, for that matter) but I knew I wanted to find a way to fight injustice and work with others to effect change. Part of that desire came from somewhere within, I suppose, but part of it came from watching the news with my dad, growing up under the fear of nuclear war under the Reagan administration, trying to make sense of the Ethiopian famine, and other world events that left me feeling unsettled throughout my childhood. In college, I finally settled on a major in environmental studies. I liked being outside, learning from/with nature, but found the "hard" sciences lacking in terms of a potential for sociopolitical engagement. As an adult, I explored a variety of pathways to bring together my multiple interests and commitments: outdoor environmental education, community organizing – locally and internationally, classroom science/ESOL teaching in Title 1 schools, teaching abroad, etc. Through all of those experiences, I came to learn firsthand about "all the lies my teachers told me" in my K-12 education. I had been socialized through schooling and family to believe in the dream of upward mobility, but in my early adulthood I became brutally aware of how that dream was only accessible to a few – and, worse, often came AT THE EXPENSE of the social, economic, political, and emotional welfare of marginalized others. For example, while living in Guatemala, I learned about the CIA takeover of the popularly elected president in Guatemala in order to secure economic gains for United Fruit (i.e., Schlesinger and Kinzer 1999), leading to a violent and oppressive dictatorship that endured for decades. My experiences teaching science in the South Bronx with no lab rooms due to overcrowding and only a classroom set of books – and making half the salary that starting teachers in the suburbs just outside of NYC made – revealed how racial and economic oppression are constituted through material-discursive practices such as the disparate funding of education through property taxes. I guess all of these things explain why I am where I am now. I think being an academic is being in a position of power. I think science and science education have that potential, too, though our field has been largely politically conservative and disengaged from justice work. I remember seeing my name outside my office when I got my first academic position as a professor at The University of Arizona. And it had this profound effect on me. I remember thinking, this is a position of power. And "[w]ith great power, there must also come great responsibility" (See Spiderman, Marvel Unlimited 1962). But I also think its potential for resistance as power is limited to – or really rests with – a collective engagement, a solidarity. And that is what is lacking in science education, though I do feel this is changing.

Alberto: It is very interesting to note how Sara's and Alexa's life paths intersect with mine. Even though we come from such different sociocultural, socioeconomic and gendered spaces, our desire to work for social justice is deeply rooted in our own encounters with struggle. For my part, I am a walking contradiction of the often cited (and often misused) statistics about the Other. I am a 'statistic' many claim to know, but know little about. I was born in a South American country, and after my mother and father divorced, my siblings and I entered the desperate world of poverty and struggle that especially afflicts single mothers anywhere. Instead of quitting high school and looking for a job as it is often the case in these self-perpetuating loops of poverty, I felt that a way out for me and my family would be to finish high school and go on to university. The promise of attaining a university degree and finding a good job seems to resonate across boundaries everywhere, and these two goals gave everyone in my family some hope as well. However, the last year of high school was the most challenging. Our limited source of income became even less certain, and some days hunger threatened to be the most dream-crushing and demoralizing force almost making me quit school.

With the kindness and encouragement of high school friends, I finished school and eventually got a scholarship to study biology abroad. At 18 years of age, Canada was to be my first trip ever outside my home country, and the beginning of my journey against racism and discrimination. It was in a small city in southern Ontario, Canada that I discovered that my skin was dark, and that made me an object of hate by some. A pair of Anglo males, not much older than me, sat behind me in a public bus and kept saying in an angry tone, "hey, Paki, go home; you're not wanted here." I nervously looked around because, being new to Canada and just beginning to learn English, I had no idea what 'Paki' meant and why those boys appeared so angry. I later found out that "Paki" was short from Pakistanis, and my dark skin and features were all those boys needed to unleash their hate just because they could.

In retrospect, after four decades, I never expected that the young Latino who sought a better education to help his single mother and siblings get out of the cycle of poverty, would become a Latin@, new immigrant, science teacher, single father; and many years later an endowed chair professor in science education. While my professional and academic locations have changed, it is unfortunate to note that my social justice work – and the need for this kind of work – seems to be more desperate. That is, being Latin@; using English as an additional language; having dark skin and speaking with an accent; coming from a single mother home background, and being a new immigrant still – after four decades – continue to represent deeply entrenched sociocultural and institutional obstacles for individuals. But worst of all, what we do know from a broad research base on these well documented obstacles is not being used to allocate the significant resources needed, nor used to provide the support required to counter these factors (Rodriguez 2004, 2015).

Throughout my career, students, colleagues and other critics have made statements like "well, look at you, aren't you living proof that the American dream works?" No, I'm not. Meritocracy is a myth instituted by those in power to deny the equitable distribution of power and resources and to avoid taking responsibility (Rodriguez 1998). I have had to work harder and continue to work harder than my peers for everything that I have accomplished, and I continue to be an object of hate and discrimination to this day. The meritocracy myth is often used to mask the many Others who "fell through the cracks," and this myth must be dismantled.

As Sara pointed out earlier, we must take on the responsibility that arises from our privileged intellectual and socioeconomic locations to purposely seek to effect sociotransformative change. And a good place to start is by acknowledging our own positionality – to reflect on how positionality influences our work, with whom we choose to work, and whose interests are being met through our work.

Alexa: You each stated so eloquently what I feel is the major driver of my own work – that these "deeply entrenched sociocultural and institutional obstacles," including privilege, meritocracy, the american dream, and white supremacy, create and maintain social (re)production (Anyon 2009). This brings us to our second topic: *public*. As Sara noted, we have great power within these academic positions, and as we examine that power, we can (and all researchers must) examine this power in relation to the folks we work with. By using the word *public* here, we are explicitly calling attention to the fact that education is a public endeavor and when we conduct research within schools, we must acknowledge and explore the ways that we are accountable to the public we serve. Are we merely getting grants and publishing papers and doing all the things that academics do for the purpose of adding to our own resumes? Or do we prioritize the educational needs of our communities within our research? It would be highly disingenuous for an education school at a research intensive university to make little or no attempt to meet the most persistent concerns in schools and the educational needs of students and teachers. The term *public* invokes the need to raise the standards that must be met for our scholarship. We urge the science education community to make research and scholarship *public* – more accountable to our communities, and particularly to the most disenfranchised communities and the most entrenched problems in science education, and more responsive to the needs of the public and participants who can benefit from it.

Returning to the concern Alberto raised, I ask us to consider: *What is your responsibility as a researcher to the communities in which you work and to the broader public, and how does your positionality influence this?* I think of my position as a researcher as someone who engages in *research with* participants. We engage as *researchers with* when we are not attempting to extract data from people but are engaged in learning, researching, and relating with them. In my work, I have noticed that there are many layers to my position and to considering who benefits from my work or how my work might impact the research participants or other teachers and students more broadly. These issues evolve within different stages of the research. As a cisgender straight white woman, I may easily be granted insider status within a school – I can pass as a person who looks like a teacher (white and female) in most schools. Yet in working with historically disenfranchised youth, I enter the classroom and our interviews as an outsider. I spend a lot of time building relationships with youth to develop trust – through being there, listening, not evaluating or judging their ideas, and creating spaces where student voice is taken seriously. This is one layer to developing the kinds of *researching with* relationships that happen in the everyday interactions on the ground when collecting data. Another layer occurs through sharing data analysis. I share findings with teachers with whom I work so that they can utilize the findings to impact their practice. Other layers occur in telling their stories, or in sharing the research findings in ways that are also authentic to the participants and respect their intensive work in classrooms. We have a great deal of responsibility to our participants, who often give freely of their time and let us into their classrooms (or other spaces) and share their ideas and experiences with us.

Sara: I think you raise important considerations for reflecting on how we as scholars *of education*, in particular, are responsive to the needs and concerns of the public, Alexa. As a former teacher in underfunded schools and economically oppressed communities, I often felt alienated by the research when I first began my PhD. I was frustrated by what Alberto has referred to as a privileging of narratives of despair over narratives of engagement (Rodriguez 2015). I remember thinking, "Hey, um, you know it would be great if you could engage some of your social capital to help us disrupt all the things." One of the ways I now try to be responsive in my research is working through the struggle (or in the words of Cornel West (2009), striving to "give heart to intellect by being true to the funk of living", p. 6) alongside teacher and student participants, when it is appropriate. This means that I am entangled within both the problems and solutions of/to inequities in science and education, though the approach is wrought with tensions, many of them unresolvable. I have talked with friends about the feeling is of being haunted by the messiness of the work, the inevitable "betrayal situations" (see Stacey 1991) of a more engaged approach. Maria Puig de la Bellcasa (2012) talks about caring not as a "feel good" warm and fuzzy state, or way of interacting with others, but rather as "material engagement in labours to sustain interdependent worlds, labours that are often associated with exploitation and domination" (p. 198).

I strive to include teacher and youth participants as collaborators in the research, so that they hold me accountable to the highest standards for ethics in representation, and have opportunities to be publicly acknowledged, if they choose, for their incredible commitment and dedication to the justice issues we try and take on in science classrooms (See Tolbert et al. 2018). Yet, these entangled/interdependent existences bring "inescapable troubles" (Puig de la Bellacasa 2012), where we may find ourselves just as befuddled as our research participants, learning with, not having the answers – when we are expected to have them, when at times we are expected to lead, etc. I remember one case where we really took an interest in the way a particular student, who had dropped out of school, who had been identified as "at-risk", who had been repeatedly moved between parents and states, had seemed to transform so dramatically from seeing science as irrelevant, boring, hard, beyond her capabilities, to then viewing science it as socially relevant, an important tool for positive change, etc. Her school attendance improved, her overall attitude in science class improved. Yet, the next year she dropped out of school. Just as science is no savior, neither are we.

The question of public also gets troubled in terms of whether/how findings are represented and reported, by whom, and to which audiences. In a racist/nativist state like Arizona, I feel like I have to tread cautiously along these lines. I think about how Eve Tuck and K. Wayne Yang's (2014) concept of ethnographic refusal comes into play here – sometimes research is not the intervention that is needed, and what is shared from a research project must be transparent and negotiated with research sites and participating communities. These are very real issues in Arizona, particularly since the passing of HB 2281 (the ethnic studies ban) and SB 1070. For example, a participatory school-based research project publishing findings that could be interpreted by the state legislature as "promoting resentment toward a race or a class of people", could have serious and deleterious consequences for the school, students, and teacher. We also had a case where we wanted to bring along a student presenter who was undocumented. In the end, her family was concerned (understandably so) about her flying to the conference given the presence of Border Patrol at the local airport and so she could not join us. While I advocate strongly for civil disobedience, I also know that I am, as an academic and a middle class White woman, more insulated (at least currently) from any fall out of our work, even as a pre-tenure faculty member. So if anyone takes the hit, I want it to be me. In practice, it is all very messy.

Alberto: Yes, it is very messy, indeed, and so contradictory! As a teacher, and later as a graduate student, I got tired really quickly of having my "consciousness raised" as others (mainly Anglo and privileged males) were calling for equality without being actually engaged in that struggle, in the trenches with the Other. I look for specific examples, but I mainly found *narratives of despair* that tended to focus on deficit frameworks to describe what was lacking or not working (often blaming the victims or making victims where they failed to notice the significant work Others were doing to improve their own condition). At the other extreme, I often found cheery narratives – those unbelievable narratives of research interventions that worked so well for everyone that their findings failed to ring true to my own experiences as a Latin@ teacher, science educator and researcher who have worked with teachers in multiple contexts. Thus, I preferred to engage, instead in *narratives of engagement* – an honest account of the challenges and successes we encounter as we seek to effect sociotransformative change in our schools. For example, in another manuscript (Rodriguez 2015), we share Gary's journey from preservice teacher through his first 2 years of teaching science at a mostly Latin@ and economically impoverished high school. While Gary (an Anglo male, former scientist) was committed to addressing equity issues, he encountered a multitude of domesticating forces that sought to mold him into the prevalent culture of low expectations. Through dialogic conversations and on-site support, Gary began to chart his own path and implemented a series of pedagogical, curricular, and transformative strategies that resulted in increased student participation and achievement. However, in the published manuscript (Rodriguez 2015), we chose not to focus only on the obstacles (narrative despair) or only on the success of the project (cheery narrative). Instead, we describe narratives of engagement that take into account the importance of honestly documenting the struggles and successes as part of the whole story. Sharing these more balanced stories will provide more practical insights for engaging in transformative work. It is about developing a toolkit of strength, commitment, perseverance, alliances, and strategies that enables us to engage with the next obstacle – and there will be more, as well as more successes.

It is at this junction that I strongly believe that we can find a common space with the Other, teachers, parents, administrators and other researchers, and where the public domain of our research can ring true to those who we are meaning to impact with our work. We must seek to move away from narratives of despair and cheery narratives and share openly the obstacles we encounter; the steps we take to address them; what we learn as research participants; and how our research participants found these steps transformative or not. This approach would take a different kind of researcher-participant relationship. One in which the researcher is not entering the research enterprise with all the answers, or with the intention to "fix" the Other, but instead entering the research enterprise with the honest desire to learn with the Other, and to recognize that the Other may be already engaged in significant struggle for change (Rodriguez 2016). This approach will make the public domain of our work more meaningful, inspiring and reach a broader audience. The pervasive inequalities we continue to observe between the haves and have nots provide compelling evidence that the public domain of our research must change, and that we must take more purposeful steps to make the fruits of our work more relevant to the very people whose lives we recount mainly to each other in academic journals and research conferences.

Alexa: In closing, all three of us are affirming that everything we do in the research process is influenced by who we are – from study design and our ethics that configure our design to interactions with folks in schools; to how we position ourselves and the results of our research; to the ethical presentation of the research. In our journal article on this topic (Tolbert et al. 2018), we contend that really rigorous research is research that does what we all have described – research that is critical, participatory, and responsive to the most salient equity issues in science education and that uses research as a tool for transformation in the lives of the students and teachers with whom we work.

Looking forward, we need to come together more and more as a community to engage in the types of critical praxis we have described here. We have a lot of work to do as a community to transform what we count as rigorous research into research that is in tune with the needs for social and personal transformation in science education. This issue can be complicated by the fact that it is tied to many students' and professors' advancement and tenure. We want dominant structures to change – that is, for funding agencies or journal editors to include our vision of rigor in their assessment of rigorous research. In the meantime, it is equally important to move forward with our own assessments of our work and to strengthen our vision of rigor through collaborative engagement within the critical science education community we are building (e.g. specifically within the SEEDS organization: Science Educators for Equity, Diversity, and Social Justice). I am heartened by the efforts we are making and by what I am learning from and alongside other educators, students, and researchers!

Sara: I think hope is in our ability to see, and enact (or, rather, co-constitute, intra-act-ing with/among), those possibilities – in solidarity with our multiple communities of justice. Pursuing those possibilities, or "lines of flight" (see Bazzul et al. in press), from what seem to be very rigid, entrenched, and/or oppressive material/discursive conditions is messy, *inevitably rigorous*, IMMEDIATELY necessary, and **morally imperative**, particularly given the recent turn of political events. I am grateful for the opportunity to engage in this important conversation with you both.

We end here with some overarching and continuing questions we pose to ourselves and others as we continue to work through these issues individually and collectively:

- Whose interests are served by our research and how might our work transform oppressive structures related to privilege and positionality?
- How do we thoughtfully and purposefully engage with participants while attending to complex issues of positioning, positionality and status?
- How do we engage in research (from praxis to publication and dissemination) that serves the purposes of social transformation in and through science education?

References

Anyon, J. (2009). *Theory and educational research: Toward critical social explanation*. New York: Routledge.

Bazzul, J., Tolbert, S., & Kayumova, S. (in press). New materialisms and science classrooms: Diagramming ontologies and critical assemblies. In K. Scantlebury & C. Milne (Eds.), *Material practice and materiality: Too long ignored in science education*. Netherlands: Springer.

Marvel Comics. (1962). Amazing Fantasy #15.

Puig de la Bellacasa, M. (2012). 'Nothing comes without its world': Thinking with care. *The Sociological Review, 60*(2), 197–216. https://doi.org/10.1111/j.1467-954X.2012.02070.x.

Rodriguez, A. J. (1998). Busting open the meritocracy myth: Rethinking equity and student achievement in science. *Journal of Women and Minorities in Science and Engineering, 4*(2&3), 195–216. https://doi.org/10.1615/JWomenMinorScienEng.v4.i2-3.80.

Rodriguez, A. J. (2004). *Turning despondency into hope: Charting new paths to improve students' achievement and participation in science education*. Tallahassee: Southeast Eisenhower Regional Consortium.

Rodriguez, A. J. (2015). Managing institutional and social challenges through sociotransformative constructivism: A longitudinal case study of a high school science teacher. *Journal of Research in Science Teaching, 52*(4), 448–460. https://doi.org/10.1002/tea.21207.

Rodriguez, A. J. (2016). For whom do we do equity and social justice work? Recasting the discourse about the other to effect transformative change. In N. M. Joseph, C. Haynes, & F. Cobb (Eds.), *Interrogating whiteness and relinquishing power: White faculty's commitment to racial consciousness in STEM education*. New York: Peter Lang. https://doi.org/10.3726/978-1-4539-1716-9.

Schlesinger, S., & Kinzer, S. (1999). *Bitter fruit: The story of the American coup in Guatemala* (2nd ed.). Cambridge, MA: Harvard University Press.

Stacey, J. (1991). Can there be a feminist ethnography? In S. B. Gluck & D. Patai (Eds.), *Women's words: The feminist practice of oral history* (pp. 111–119). New York: Routledge. https://doi.org/10.1093/ohr/20.1.105.

Tolbert, S., Schindel, A., Rodriguez, A. J. (2018). Relevance and relational responsibility in justice-oriented science education research. *Science Education, 102*(4), 796–819. https://doi.org/10.1002/sce.21446

Tuck, E., & Yang, K. W. (2014). R words: Refusing research. In D. Paris & M. Winn (Eds.), *Humanizing research: Decolonizing qualitative inquiry with youth and communities* (pp. 223–248). Los Angeles: Sage.

West, C. (2009). *Brother west: Living and loving out loud, a memoir*. New York: SmileyBooks.

Alexandra Schindel is an Assistant Professor in the Graduate School of Education at the University at Buffalo, SUNY. Alexa researches the experiences of high school-aged youth and their teachers when engaged in socially just science learning and teaching in schools serving minoritzed students. In collaboration with Sara Tolbert, Alexa studies teacher-initiated social justice activities in formal school science. One of their current projects involves working with five middle and high school science teachers across the U.S. to identify and reflect on "critical incidents" of student empowerment in their science classrooms. Another facet of Alexa's research investigates teaching and learning in environmental science education and consider youth's civic environmental engagement. Her most recent publications focus on themes of empowerment, sustainability, critical care, and critical pedagogy of place.

Sara Tolbert is Associate Professor in the Teaching, Learning, & Sociocultural Studies Department at the University of Arizona's College of Education, and a former secondary science, ESOL, and environmental education teacher. Sara Tolbert and Alexa Schindel partner with secondary science teachers across a variety of teaching contexts to better understand how school science can be a vehicle for youth empowerment, with particular attention to how science education can engage students in local and global justice issues. Sara has also collaborated on the design and implementation of new transformative models for science teacher education including those that support and honor the experiences of marginalized Indigenous, refugee, immigrant, and(or) emergent bilingual students in secondary science. In some of her most recent work, Sara contemplates aesthetics, science, politics, art, activisms, traditional indigenous lifeways, and string figures/speculative futures as/for acts of love and collective survivance.

Alberto J. Rodriguez is the Mary Endres Chair in Elementary Education and Professor of Cross-Cultural Science Education in the Department of Curriculum and Instruction at Purdue University. His research focuses on the use of sociotransformative constructivism (sTc) as a theoretical framework that merges critical cross-cultural education tenets (as a theory of social justice) with social constructivism (as a theory of learning). Thus, Dr. Rodriguez is investigating how teachers can make their pedagogy and curriculum more culturally and socially relevant to all students, as well as how teachers can better integrate STEM across all curriculum subjects.

Currently, Dr. Rodriguez is the PI of the 20/20 Vision for Transdisciplinary Cross-Cultural STEM Project. This study brings together teacher education faculty from across all the curriculum areas to collaborate in the design of cross-cultural and socially relevant integrated STEM modules. Dr. Rodriguez's work has been published in various journals such as, the American Educational Research Journal, the Journal of Research in Science Teaching, Research in Science Education, the Journal of Teacher Education, Theory into Practice, and many others. One of his previously published articles was selected for the *Multicultural Science Education, Equity and Social Justice* special issue of the Journal of Research in Science Teaching (JRST). For this special issue, 9 of the most influential science education articles in the previous 30 years were selected for re-print. The selected article was: Strategies for counterresistance: *Toward sociotransformative constructivism and learning to teach science for diversity and for understanding* (reprinted in JRST, November 2011). Dr. Rodriguez has also edited and co-edited four research-based books, and one co-edited volume is in-press. The co-edited volume with Rick Kitchen (math education) entitled, *Preparing Prospective Mathematics and Science Teachers to Teach for Diversity: Promising Strategies for Transformative Action* (2005), was selected as an Outstanding Academic Title in 2005 by Choice Magazine.

Dr. Rodriguez recently received the *Innovations in Research on Diversity in Teacher Education Award* from the American Educational Research Association (AERA), Division K (Teaching and Teacher Education, 2017). He also received the *Kappa Delta Pi – Teaching and Teacher Education Research Award* from AERA in 2000, and the New Mexico State University's *Award for Exceptional Achievements in Creative Scholarly Activity* in 2002.

Chapter 19
Playing Within/Against Entombed Scholarship: Episodes in an Academic Life

Noel Gough

19.1 Prologue

Reflecting on writing as a method of inquiry, Laurel Richardson (2001) insists that she writes

"to learn something that I did not know before I wrote it. I was taught, though… not to write until I knew what I wanted to say, until my points were organized and outlined. No surprise, this static writing model coheres with mechanistic scientism… and entombed scholarship" (p. 35). This essay recounts two episodes in my career as an academic science educator in which I deliberately resist entombed scholarship.

The first episode is a lightly edited restatement of my Introduction to *Laboratories in Fiction* (Gough 1993), a monograph commissioned by Deakin University to resource a new postgraduate science education course. I describe the genealogy of *Laboratories in Fiction* elsewhere (Gough 2015), and John Weaver (1999) provides a generous appraisal of its significance, but the monograph itself is long out of print. The introduction exhibits my preference for first-person voice and foreshadows my nascent disposition to perform educational inquiries as narrative experiments catalyzed by intertextual readings of popular media. Borrowing Deleuze and Guattari's (1987) concept of *rhizome*, I characterize these experiments as *rhizosemiotic play* (Gough 2007), which I use to explore representations of gravity and malaria in textbooks and news media (Gough 2006) and to generate connections between thought experiments in scientific and educational inquiry (Gough 2010).

The second episode mobilizes popular culture's zombie imaginary to identify dead ideas that "walk" among science educators, and is little more than a rant arising from my efforts to specify a curriculum for biopolitical literacy in science teacher education (Gough 2017).

N. Gough (✉)
Latrobe University, Melbourne, Australia
e-mail: N.Gough@latrobe.edu.au

J. Bazzul, C. Siry (eds.), *Critical Voices in Science Education Research*,
Cultural Studies of Science Education 17,
https://doi.org/10.1007/978-3-319-99990-6_19

19.2 Episode 1 (1993): Introducing Laboratories in Fiction

I encountered no "science" in primary school (although I recall "nature studies"). My induction into Science began in secondary school during the late 1950s, when it was easy to be optimistic about science and technology. Intense sunspot activity produced conditions that made the aurora australis visible from southern Australia and local media coverage of the 1957–1958 International Geophysical Year (IGY) highlighted atmospheric physics research by Australian scientists in Antarctica. I recall watching the aurora's luminescent flares pulsating above the horizon near my home in South Melbourne and my thoughts often drifted towards imagining myself as an Antarctic scientist. But my heroic visions and scientific optimism were not drawn from science textbooks or journalism but from novels such as Ivan Southall's (1956) *Simon Black in the Antarctic*, one in his series for young people featuring the eponymous Simon Black, a brilliant aerospace inventor, engineer, pilot and United Nations special agent whose adventures took him to exotic locations (including Mars and Venus). In 1957 I watched Sputnik 1 orbiting the earth, delighted to see – *at last!* – tangible evidence of humankind's entry into the space age (I was oblivious to its Cold War implications). Recalling the late 1950s, Donald Fagen (1982) captures the buoyant mood of many young people in his song, "IGY (International Geophysical Year)":

Get your ticket to that wheel in space
While there's time…
You'll be a witness to that game of chance in the sky
The fix is in
You know we've got to win
Here at home we'll play in the city
Powered by the sun
Perfect weather for a streamlined world
There'll be spandex jackets one for everyone

… A just machine to make big decisions
Programmed by fellows with compassion and vision
We'll be clean when their work is done
We'll be eternally free yes and eternally young
What a beautiful world this'll be
What a glorious time to be free

At high school, then throughout my undergraduate studies in science and education and my years as a high school science teacher, my enthusiasm for science never waned, although I became more cynical about the "compassion and vision" of those who made "big decisions." When I became a science teacher educator in 1972 I began to reflect critically on the assumptions underlying my enthusiasm. Given the length of time I endured academia, and to pre-empt any tendencies towards self-satisfaction or complacency, I cannot resist quoting Leonard Cohen (1988):

They sentenced me to twenty years of boredom
For trying to change the system from within
I'm coming now I'm coming to reward them
First we take Manhattan, then we take Berlin

Laboratories in Fiction attempts "to change the system from within" and illustrates one way in which I tried to escape "boredom." I interpret "coming to reward them" as dedicating this monograph to the now absent presences of my high school and university science teachers: it could only have been written without them.

Other verses of Cohen's song are more directly pertinent to the content of this monograph:

I'm guided by a signal from the heavens
I'm guided by this birthmark on my skin
I'm guided by the beauty of our weapons
First we take Manhattan, then we take Berlin...

I don't like your fashion business, mister
I don't like these drugs that keep you thin
I don't like what happened to my sister
First we take Manhattan, then we take Berlin

Cohen's compressed images depict some of the confusions and contradictions that attend the ambiguous roles of science and technology in shaping the late twentieth century world. The ironies merely whispered in Fagen's evocation of the late 1950s ("The fix is in…we've got to win") are an ominous presence in the late 1980s. Anticipating a "wheel in space" – like the space station that orbits the Earth to the tune of the "Blue Danube" waltz in Stanley Kubrick's (1968) *2001: A Space Odyssey*– is a very different "signal from the heavens" from what was envisaged by the Strategic Defense Initiative (SDI) announced by US President Ronald Reagan in 1983. The mainstream media nick-named the SDI, "Star Wars," deceptively cloaking it in the grandeur of George Lucas's (1977) cinematic saga. But contemplating "the beauty of our weapons" in a mythic realm (*Star Wars* is set "a long time ago in a galaxy far away") is significantly different from romanticizing them in the world *we* inhabit (when I wrote *Laboratories in Fiction*, the SDI was intended to be here, now).

Similarly, the world that Fagen suggests we once anticipated – affluent ("spandex jackets… for everyone"), hygienic ("we'll be clean") and healthy ("we'll be… eternally young") – did not emerge from the work of "fellows with compassion and vision." Rather, as Cohen insinuates, science and technology added to the power of capitalist and patriarchal interests to exploit and oppress people, especially women, through the global fashion industry, multinational drug corporations and experimentation in reproductive technologies. Cohen sings, "I don't like what happened to my sister," and I don't like it either, which is why we need to "take" Manhattan and Berlin – capture and critically mobilize the cultural resources of these emblematic sites of Western society so central to the art and commerce of their respective continents, Manhattan and Berlin are culturally connected via the development of the USA's atomic bomb (aka The Manhattan Project) as a precursor to the Eastern bloc erecting the Berlin wall and maintaining it for nearly two decades as a symbol and materialization of the Cold War.

I interpret "First we take Manhattan" as a series of snapshots sampling cultural manifestations of postmodernity, which Katherine Hayles (1990) characterizes as the "convoluted ambiguity" accompanying "the realization that what has always been thought of as the essential, unvarying components of human experience are not natural facts of life but social constructions," (p. 265) an ambiguity revealed in the diverse – and not necessarily welcome – products of science and technology ("the beauty of our weapons," "the drugs that keep you thin") and the power arrangements that mediate their uses. Technologically sophisticated weapons and drugs, and the global marketplace controlling them, have reconstructed our "natural" senses of beauty and health in complex and contradictory ways.

Juxtaposing "IGY" with "First we take Manhattan" problematizes convoluted interrelationships of science, technology and society in ways that are also addressed in various academic disciplines, with feminist scholars providing some of the most cogent and trenchant critiques. For example, Ruth Bleier (1988) writes:

> Science is an integral part, expression, and product of a culture's complex set of ideologies, and it has ideological commitments to certain social beliefs, values, and goals. These commitments are, on the one hand, a source of its great strength and value and, on the other, the source of its oppressive power… It was, after all, in response to our society's social beliefs, values, and urgent needs that scientists, for example, worked to develop antibiotics before and during the Second World War, at the same time that other scientists worked to develop the atom bomb, a weapon designed not to save lives by bringing a quick end to the war with Japan but to announce the ultimate phallic power and hegemony of United States capitalism in the leadership of the coming war against the Soviet Union. (p. 57)

Like Cohen, Bleier uses images of medicine and weapons to invoke some of the dilemmas we face when trying to understand science and transform its terrifying power. That is, "IGY" and "First we take Manhattan" can be interpreted as texts that complement Bleier's essay – different but convergent expressions of the hopes and fears aroused by the promises and threats of scientific "progress."

You do not need to agree with my idiosyncratic interpretation of "First we take Manhattan" to acknowledge that it generates meanings appropriate to studies of science and technology. Popular art is full of malleable allusions that can be retrofitted to the consumer's consciousness. Some observers see popular media as ephemeral and/or disposable. But popular artifacts – snippets of song lyrics, archetypal images, pithy lines of dialogue, characters in movies, TV shows, novels, plays and comic strips – readily work themselves into our individual and collective memories and mythologies. As J.G. Ballard (quoted in Vale and Juno 1984) observes, "pop artists deal with the lowly trivia of possessions and equipment that the present generation is lugging along with it on its safari into the future" (p. 155). I did not quote "First we take Manhattan" just because it speaks to me of issues in science and technology studies, but also because Cohen's songs are among the "lowly trivia of possessions" I have been "lugging… into the future" since 1966. They are items of "equipment" that connect me with the world and helps me to make sense of it.

Laboratories in Fiction makes connections between science and popular media that enrich science education and respect popular art and artists. Consider, for example, the following excerpt from Janette Turner Hospital's (1988) novel, *Charades*:

> "Question," Charade says. "If a woman stands in the middle of Massachusetts Avenue facing MIT, but her memory is so vividly snagged on one particular day of her childhood in the village of Le Raincy that she is unaware… that she is *oblivious* to the cars around her and so she is hit, run over, killed… Is she more truly in Boston or France when she dies?"
>
> "Well put," Koenig says. "The indeterminacy problem in a nutshell." (p. 191)

This passage relates to science education in several ways. Firstly, it illustrates one way in which meanings emerge, unforced, in the course of everyday conversation. Charade and Koenig are not involved in a didactic exchange in which one is trying to transmit to the other a stipulative definition of quantum indeterminacy – they are simply having a chat. Secondly, given that Charade's question is itself a response to something that Koenig has said previously (the details of which are unimportant), both characters are modeling a strategy for good teaching that Garth Boomer (1982) terms "connecting":

> The teacher is a senior reader of the school culture and special senior reader of the specialist subculture of the subject. Wittingly or unwittingly, he/she is demonstrating how to be a reader and maker of meaning… The more richly the teacher can spin a tapestry of metaphor and analogy into a "thick" redundant text of thinking about something new, the more likely it is that students will find a way in. If students are encouraged to spin out reciprocally their own webs of anecdote, metaphor and analogy, it is less likely that some will remain outside the next text. The art of generating apt analogy and metaphor is central to the "reading" teacher's task. (pp. 119–120)

Charade and Koenig "read" each other's speech acts and respond by spinning reciprocally "webs of anecdote, metaphor and analogy." Their conversation also exemplifies a point of entry into science subject matter that is different from that used by most science teachers. *Charades* is a popular novel that incorporates ideas drawn from quantum mechanics and other aspects of subatomic physics. In part this is because one key character is a research physicist but is also because the author is playing creatively with the existential and metaphysical implications of quantum theory. *Charades* is not only more pleasurable to read than most physics textbooks, but also situates meanings drawn from the subatomic world within the politics of everyday human activity and experience. I also note that the majority of contemporary school science textbooks ignore the physics *Charades* explores. Although the "new physics" has been with us since the late nineteenth century, few late twentieth century textbooks pay more than lip service to its existence, let alone explore its implications for understanding human experience.

Neither *Charades* nor "First we take Manhattan" are isolated examples. I could have made similar points using Tom Robbins' (1990) novel *Skinny Legs and All* and Paul Simon's (1986) song "The boy in the bubble." *Laboratories in Fiction* demonstrates that popular media – music, movies, comics, novels, and other media popular with children and young adults – are rich, meaningful sources of information, images and insights concerning science, technology and society and their interrelationships. I also argue that popular media are much more than "icing on the cake," a way of illustrating subject matters of science in ways that are "palatable" to young people. Rather, popular media provide *sites for inquiries* into the meanings of scientific concepts and methods and provide some "equipment" (in Ballard's sense) for investigating problems and issues of science, technology and society.

I thus affirm for science education a position adopted already by many teachers in subjects such as English language and literature, media studies and social education, namely, that popular media are "texts" in their own right and merit close "reading," and critical and creative responses, from both teachers and learners. I chose the title, *Laboratories in Fiction*, with the deliberate intention of emphasizing and exemplifying two propositions:

- Laboratories, in their various roles as sites, symbols, emblems and metaphors of scientific labor, are represented in numerous and diverse ways in popular media, and these images of science can be generative foci of science education.
- Many examples of popular media are "laboratories of ideas" in which *meanings* are subjected to experimentation.

I use the term "popular" to mean media produced with the deliberate intention of having wide appeal, especially (but not exclusively) to young people. It is not necessary for a work to *achieve* wide appeal to be designated "popular" – a pop song remains a pop song even if it fails to appear in the *Billboard* Hot 100. I also focus chiefly on works of "art" (such as songs) and "fiction" (such as comics, novels, movies) rather than what I prefer to call *science journalism* – although magazines like *New Scientist* and many TV documentaries are clearly intended to be "popular."

Marshall McLuhan (1960) argues persuasively for teaching and learning *with* the texts and artifacts of mass media and popular culture:

> It's misleading to suppose there's any basic difference between education and entertainment. This distinction merely relieves people of the responsibility of looking into the matter. It's like setting up a distinction between didactic and lyric poetry on the grounds that one teaches, the other pleases. However, it's always been true that whatever pleases teaches more effectively. Where student interest is already focused is the natural point at which to be in the elucidation of other problems and interests. The educational task is not only to provide basic tools of perception but also to develop judgment and discrimination with ordinary social experience… To be articulate and discriminating about ordinary affairs and information is the mark of an educated [person]. (p. 3)

That learning should begin "where student interest is already focused" is a familiar pedagogical platitude, but I suspect that it might have been manifested more in well-intentioned teachers' aspirations than in young people's lived experience. If teachers are to develop "judgment and discrimination with ordinary social experience," they cannot merely *begin* "where student interest is already focused" and then retreat to the relative security of their own interests and experience. Boomer's (1982) notion of teaching as "bushcraft" is pertinent:

> In the ecology of the school "bush" there is a bewildering array of texts, tests, assignments and artefacts. The teacher should be used to finding interesting and pertinent specimens and talking about their characteristics, habits and habitats. Students should be encouraged to familiarize themselves with funny creatures like science textbooks, learning how to tame them, remembering where dangers lurk... Teachers should not drive students in a tourist bus through the school curriculum, encouraging the bland recital of tourist blurbs. Students should be obliged to savour the texture of life, wild and rich. (p. 119)

I agree with Boomer in most respects (*Laboratories in Fiction* is itself a guide to taming "funny creatures like science textbooks"), but I would add that *teachers* also should be "obliged to savour" the wild and rich textures of students' lives. I concur with Ballard's (1985) interpretation of the ways we experience our worlds:

> The most prudent and effective method of dealing with the world around us is to assume that it is a complete fiction… We live in a world ruled by fictions of every kind – mass-merchandizing, advertising, politics conducted as a branch of advertising, the instant translation of science and technology into popular imagery, the increasing blurring and intermingling of identities within the realm of consumer goods…We live inside an enormous novel. (p. 8)

I understand "fiction," as derived from the Latin *fictio*, as something fashioned by a human agent, and thus interpret Ballard's reference to living "inside an enormous novel" as one figuration of living "in a world ruled by fictions of every kind." Although many of us live inside enormous novels (the plural is important), the subjectivities of many young people reside (at least partially) in the enormous fictions produced by multi-media franchises that market videos, movies, games and other merchandise. Teachers must be open to and capable of engaging empathetically and constructively with these fictional worlds.

19.3 Interlude

During the quarter century that has elapsed since I wrote *Laboratories in Fiction*, there have been both continuities and changes in the objects of my inquiries and the research methodologies I privilege. I still focus on science and environmental education, but transnational curriculum inquiry and the politics of complexity reduction are increasingly salient research interests. My continuing disposition to deploy methodologies informed by narrative and poststructuralist theorizing has been refined by reference to posthumanism and my preference for what I prefer to call a *postparadigmatic* position (Gough 2016). I also try to honor the spirit of Deleuze's (1995) encouragement for "writing to… free life from where it's trapped" (pp. 140–1). What follows is an attempt to free science educators from the traps of dead ideas.

19.4 Episode 2 (2017): Dead Ideas That Walk Among Science Educators; an Incomplete List

Lyn Carter's (2014) essay on science education and neoliberalism links Foucault's late 1970s lectures on biopolitics with the September 2011 Occupy Wall Street protests. Carter finds it "difficult to come to grips with…Occupy":

> There is so much one could interrogate – from the protester demographics of the mainly highly-educated young White males and the concomitant elision and erasure of the racialised nature of inequality, to the information-age protest style with its own generator, YouTube™ videos, tweets, blog posts and help from hacktivist group *Anonymous*. (p. 30)

Carter does not mention that many of the Occupy protesters expressed their political discontent by dressing as zombies (Daily Mail Reporter 2011). Several recent critiques of neoliberal politics and economics foreground the zombie imaginary, including John Quiggin's (2010) *Zombie Economics: How Dead Ideas Still Walk Among Us,* David McNally's (2011) *Monsters of the Market*, and Chris Harman's (2012) *Zombie Capitalism.*

I borrow Quiggin's subtitle to ask: how *do* dead ideas still walk among us science educators? I begin here a list of such dead ideas and invite readers/colleagues to add to it (I would be delighted if someone volunteered to start a blog to which anyone interested could contribute). My list is very short because I already exceed the word limit for this chapter.

19.4.1 Dead Idea #1: School Science Laboratories

Do a Google™ search for "school science laboratories" and select the "Images" tab. You will find page after page of images that demonstrate the materialization of a dead idea. School laboratories are stereotypical gestures towards diverse sites of scientific labor. They are equipped with apparatus that fits Margaret Mead and Rhoda Metraux's (1957) image of a scientist as "a man who wears a white coat… surrounded by… test tubes, bunsen burners, flasks and bottles" (p. 386). The activities that take place in such classrooms – indeed, the activities that *can* take place in them – bear little or no resemblance to contemporary scientific practice. After World War 2 science became highly industrialized and technologized "Big Science" – big budgets, big staff, big machines –requiring very different facilities from those available in school laboratories. Also many scientific specializations – mathematical, physical, biological, cosmological, etc. – moved away from studying the *material* structures of simple systems (the prime foci of mainstream science from Newton's day until the late nineteenth century), towards modeling the *informational* structures of complex systems – protein folding in cell nuclei, task switching in bacteria colonies, far-from-equilibrium chemical reactions – through computer simulations (Casti 1997). Little of what now counts as "progress" in science is accomplished by the individualistic, small-scale, low-tech "bench work" for which school laboratories are designed.

Why do we persist? How can we resist?

19.4.2 Dead Idea #2: The "Balance" of Nature

Recent research by Nicholls and Stephenson (2015) examines teachers' personal and professional beliefs about climate change education. They analyze survey data from a large sample of primary and secondary teachers to identify teachers' understandings and beliefs about the realities, causes, and consequences of climate change, and how they conceptualize climate change education in terms of content and processes. They asked teachers to express in their own words what climate change education involves. The dominant theme that emerged was:

> *Balance or Both Sides of the Climate Change "Debate"*
> The idea of "balance" and presenting students with "both sides of the climate change argument" or a balanced perspective was most frequently identified… as important in climate change education. Teachers identified that there existed a "for and against" argument or more than one side to climate change that students should be made aware of. Not telling students what to believe but allowing them to review or be given all "sides" of the argument so they were able to "make up their own mind" about climate change was also considered important by teachers. (p. 25)

In the light of poststructuralist understandings of subjectivity and agency, I interpret these teachers' standpoints as evidence of the constitutive force of a discourse of dead ideas centred on deeply sedimented conceptions of "natural" order – order as stability, predictability, and equilibrium. Conceptions of "natural" order are pervasive in many disciplines, but I speculate that most can be attributed in to a "success" of conventional science *mis*education, namely, the "textbook ecology" received by undergraduates in US colleges and universities (and beyond) for more than 50 years.

During the post-World War II period, under Eugene Odum's leadership, the US version of systems ecology privileged the concept of the ecosystem as an enduring emblem of "natural" order, epitomised by the dominance of the "balance of nature" metaphor, which as Kim Cuddington (2001) argues, "is shorthand for a paradigmatic view of nature as a beneficent force" (p. 463). Environmental historian Donald Worster (1995) argues that Odum's (1953) textbook, *Fundamentals of Ecology*, (and its four subsequent editions) "laid so much stress on natural order that it came close to dehistoricizing nature altogether" (p. 70). He also notes that during the 1970s and 1980s "the field of ecology… demolished Eugene Odum's portrayal of a world of ecosystems tending towards equilibrium" (p. 72). Worster cites numerous studies supporting the view that the concept of the ecosystem receded in usefulness and that even the word "ecosystem" lost its former implications of order and equilibrium. Similarly, Andrew Jamison (1993) notes that systems ecology contributed very little to the solution of environmental problems and, by the late 1970s, evolutionary approaches had become increasingly popular among ecologists, "so that systems ecology today is only one … of a number of competing ecological paradigms" (p. 202). Nevertheless, successive editions of Odum's textbook helped his ideas persist as lumbering zombies of US systems ecology, a particularly appropriate metaphor given the publication of a fifth edition of *Fundamentals of Ecology* (Odum and Barrett 2005) 3 years after his death – entombed scholarship epitomized.

Gregory Cooper (2001) observes that in population and community ecology, "the balance of nature idea … has worked in the background, shaping inquiry" (p. 482), but that it has been argued largely on conceptual rather than empirical grounds. It is thus significant that Robert Ulanowicz's (2009) empirical studies of trophic exchanges, thermodynamics and causality in living systems emphasise chance, disarray and randomness as necessary conditions for emergence and autonomy in the natural world.

It is more than a little ironic that the persistence of a dead idea in the science of ecology still provides science teachers with an excuse for abrogating their professional responsibility to teach the scientific evidence for anthropogenic climate change.

How do we resist?

19.5 Epilogue

The conventional conclusion in academic writing is a component of the "static model" to which Richardson (2001) refers. I want my writing to encourage activity, not stasis.

I will close simply by encouraging readers to identify, question, critique and resist the "entombed scholarship" that continues to deaden science education. They can do this by being alert to, and deploying, the rich variety of cultural and academic resources to which they *and their students* have access, which I hope I have exemplified in the two episodes recounted above. I encourage you to advertise and add to the list of dead ideas I have commenced.

References

Ballard, J. G. (1985). *Introduction to the French edition of Crash* (pp. 5–9). London: Paladin.

Bleier, R. (1988). Lab coat: Robe of innocence or Klansman's sheet? In T. de Lauretis (Ed.), *Feminist studies/critical studies* (pp. 55–66). Basingstoke: Macmillan.

Boomer, G. (Ed.). (1982). *Negotiating the curriculum*. Gosford: Ashton Scholastic.

Carter, L. (2014). The elephant in the room: science education, neoliberalism and resistance. In L. Bencze & S. Alsop (Eds.), *Activist science and technology education* (pp. 23–35). Dordrecht: Springer. https://doi.org/10.1007/978-94-007-4360-1_2.

Casti, J. L. (1997). *Would-be worlds: How simulation is changing the frontiers of science*. New York: Wiley.

Cohen, L. (1988). *First we take Manhattan [Song from the Columbia Records album I'm Your Man]*. Lyrics retrieved 21 November 2016, from http://www.azlyrics.com/lyrics/leonardcohen/firstwetakemanhattan.html.

Cooper, G. (2001). Must there be a balance of nature? *Biology and Philosophy, 16*(4), 481–506. https://doi.org/10.1023/A:1011935220219.

Cuddington, K. (2001). The 'balance of nature' metaphor and equilibrium in population ecology. *Biology and Philosophy, 16*(4), 463–479. https://doi.org/10.1023/A:1011910014900.

Daily Mail Reporter. (2011). Wall street demonstrators dressed as 'corporate zombies' lurch past stock exchange as protests spread beyond America. Retrieved 24 April 2015 from http://www.dailymail.co.uk/news/article-2044983.

Deleuze, G. (1995). *Negotiations 1972–1990* (trans: Martin Joughin). New York: Columbia University Press.

Deleuze, G., & Guattari, F. (1987). *A thousand plateaus: Capitalism and schizophrenia* (trans: Brian Massumi). Minneapolis: University of Minnesota Press.

Fagen, D. (1982). IGY (International Geophysical Year), [Song from the Warner Bros. album *The Nightfly*]. Lyrics retrieved 21 November 2016, from http://www.oldielyrics.com/lyrics/donald_fagen/i_g_y.html.

Gough, N. (1993). *Laboratories in fiction: Science education and popular media*. Geelong: Deakin University.

Gough, N. (2006). Shaking the tree, making a rhizome: Towards a nomadic geophilosophy of science education. *Educational Philosophy and Theory, 38*(5), 625–645. https://doi.org/10.1111/j.1469-5812.2006.00216.x.

Gough, N. (2007). Changing planes: Rhizosemiotic play in transnational curriculum inquiry. *Studies in Philosophy and Education, 26*(3), 279–294. https://doi.org/10.1007/s11217-007-9034-6.

Gough, N. (2010). Performing imaginative inquiry: Narrative experiments and rhizosemiotic play. In T. W. Nielsen, R. Fitzgerald, & M. Fettes (Eds.), Imagination in educational theory and practice: A many-sided vision (pp. 42–60). Newcastle upon Tyne: Cambridge Scholars Publishing.

Gough, N. (2015). Undoing anthropocentrism in educational inquiry: A phildickian space odyssey? In N. J. Snaza & J. A. Weaver (Eds.), *Posthumanism and educational research* (pp. 151–166). New York: Routledge.

Gough, N. (2016). Postparadigmatic materialisms: A 'new movement of thought' for outdoor environmental education research? *Journal of Outdoor and Environmental Education, 19*(2), 51–65.

Gough, N. (2017). Specifying a curriculum for biopolitical critical literacy in science teacher education: Exploring roles for science fiction. *Cultural Studies of Science Education, 12*(4) 769–794. http://link.springer.com/article/10.1007/s11422-017-9834-0.

Harman, C. (2012). *Zombie capitalism: Global crisis and the relevance of Marx*. Chicago: Haymarket Books.

Hayles, N. K. (1990). *Chaos bound: Orderly disorder in contemporary literature and science*. Ithaca: Cornell University Press.

Hospital, J. T. (1988). *Charades*. St Lucia: Queensland University Press.

Jamison, A. (1993). National political cultures and the exchange of knowledge: The case of systems ecology. In E. Crawford, T. Shinn, & S. Sörlin (Eds.), *Denationalizing science: The contexts of international scientific practice* (pp. 187–208). Dordrecht: Kluwer. https://doi.org/10.1007/978-94-017-1221-7_7.

Kubrick, S. (1968). *2001: A Space Odyssey*. USA: Metro-Goldwyn-Mayer.

Lucas, G. (1977). *Star Wars*. USA: 20th Century Fox.

McLuhan, M. (1960). Classroom without walls. In E. Carpenter & M. McLuhan (Eds.), *Explorations in communication* (pp. 1–3). Boston: Beacon Press.

McNally, D. (2011). *Monsters of the market: Zombies, vampires and global capitalism*. Leiden: Brill. https://doi.org/10.1163/ej.9789004201576.i-2966.

Mead, M., & Metraux, R. (1957). The image of the scientist among high school students: A pilot study. *Science, 126*, 384–390. https://doi.org/10.1126/science.126.3270.384.

Nicholls, J., & Stephenson, R. B. (2015). Queensland teachers' understandings of education for climate change. *etropic, 14*(1), 21–27 http://www.jcu.edu.au/etropic.

Odum, E. P. (1953). *Fundamentals of Ecology*. Philadelphia PA: W.B. Saunders.

Odum, E. P., & Barrett, G. W. (2005). *Fundamentals of ecology* (5th ed.). Belmont: Thomson Brooks/Cole.

Quiggin, J. (2010). *Zombie economics: How dead ideas still walk among us*. Princeton: Princeton University Press.

Richardson, L. (2001). Getting personal: writing-stories. *International Journal of Qualitative Studies in Education, 14*(1), 33–38. https://doi.org/10.1080/09518390010007647.

Robbins, T. (1990). *Skinny legs and all*. New York: Bantam.

Simon, P. (1986). The boy in the bubble [Song from the Warner Bros. album Graceland].

Southall, I. (1956). *Simon Black in the Antarctic*. Sydney: Angus & Robertson.

Ulanowicz, R. E. (2009). *A third window: Natural life beyond Newton and Darwin*. West Conshohocken: Templeton Foundation Press.

Vale, V., & Juno, A. (Eds.). (1984). Re/Search # 8/9: J.G. Ballard. San Francisco: Re/Search Publications.

Weaver, J. A. (1999). Synthetically growing a post-human curriculum: Noel Gough's curriculum as a popular cultural text. *Journal of Curriculum Theorizing, 15*(4), 161–169.

Worster, D. (1995). Nature and the disorder of history. In M. E. Soulé & G. Lease (Eds.), *Reinventing nature? Responses to postmodern deconstruction* (pp. 65–85). Washington DC: Island Press.

Noel Gough is Professor Emeritus (and Foundation Professor of Outdoor and Environmental Education, 2006–2014) in the School of Education at La Trobe University, Melbourne, Australia. He previously held senior appointments at the University of Canberra and Deakin University and has been a visiting professor at universities in Canada, China, South Africa and the UK. His research interests include curriculum inquiry and research methodology, with particular reference to environmental and science education. He received the inaugural Australian Museum Eureka Prize for environmental education research in 1997, and is founding editor of *Transnational Curriculum Inquiry*. He is a past president (2008) and Honorary Life Member of the Australian Association for Research in Education.

Chapter 20
Multiplicitous Moments: The Inculcation, Abstraction, and Resistance to the Face of the Novice Science Teacher

Maria F. G. Wallace

At first, I was scared to question the conventions of educational research and prevailing assumptions circulating about what a science teacher educator needs to 'be like,' know and do. But more recently I have begun to cautiously embrace my ability to delve within myself – to feel what makes my stomach churn, to hear what makes me 'zone out,' to see what makes me cry and to taste the words that make me gasp. By attending to these emotive responses, driven by dominant research conventions, I have begun to embrace my ability to 'feel around' my heart, mind, and body to imagine something not (quite) yet in research on science teacher induction. As Karen Barad (2007, p. 54) asserts, "theorizing must be understood as an embodied practice," I realize I am not alone in this process. However, even in Barad's (2007) statement, I grapple with the language of 'theorizing' as it is often perceived by educators to be something intangible, abstract, and distant from one's practice as a science teacher or science teacher educator. My emerging views on science teacher education and 'doing research' are guided by the assumption that theorizing is critical to one's ability to recognize one's practice, life, and entangled meanings as an emergent process from becoming-with/in the world (Wallace, in pressA). As Dillard (2012, p. 19) states, "Everyone theorizes. It's how human beings make sense of our lives and work."

Scary theoretical moments shape my work as a becoming-science teacher educator. Throughout my doctoral education, I found myself in many tenuous stages of discomfort with myself, my research, and a once familiar reality. From dominant conceptions of what it means "to do research" to prevailing definitions of what means to "become a science teacher," I have lost *the way* (Lather 2007). Instead of engaging in my doctoral education program for the end purpose of a degree, I endeavored to navigate the implicit and explicit ways myself and others shape the world *and* get shaped by the world.

M. F. G. Wallace (✉)
Millsaps College, Jackson, MS, USA
e-mail: maria.wallace@millsaps.edu

J. Bazzul, C. Siry (eds.), *Critical Voices in Science Education Research*,
Cultural Studies of Science Education 17,
https://doi.org/10.1007/978-3-319-99990-6_20

While this chapter highlights four personal moments that diverge, rub up against, and perturb my becoming “a researcher” of science teacher induction and “science teacher educator,” the chapter also lives within an overarching series of questions I continuously (re)engage throughout my academic journey: What does it mean ‘to know’ experience? What counts as “experience”? What are the enduring implications of “knowing” someone else’s “experience”? How dare I try to write someone else’s experience, when I cannot even fully know my own? As someone who draws on feminist poststructural theory and post-human materalisms, I live in these questions, which problematize knowledge (and implicated subjects) as fixed rationalizable entities. It is from this form of criticality I share ontological tensions across instances that contextualize ways in which researchers of science teacher induction re/produce particular kinds of novice science teachers.

I am continously confused by the contradictory assumptions circulating through research on science teacher induction and education. Researchers are so often entrenched in maintaining the traditions of inquiry that the very basic ideas in which their work is grounded are rarely carefully examined. The one example I always come back to is the notion of teaching experience. Many times science teacher educators, at the pre-service level of teacher education, are advocating for ambitious science teaching practices, yet once science teachers enter formal employment, the recognition of novice science teacher experience is wiped away. Why? My academic journey has and continues to question taken-for-granted assumptions oversaturating research on science teacher induction and education. Through this non-linear engagement with scholarly inquiry, there have been many moments of criticism related to the “impact” of my work. However, more than anything these questions might create moments of productive pause among science teacher educators and researchers. Science teacher education and induction, like science education more broadly, necessitate ethicopolitical hesitations (Wallace 2018).

20.1 Navigating the Terrain of Multiplicitous Moments

The sections that follow carve out multiple moments of hesitation for myself and other researchers of science teacher induction to embrace as a productive space for thinking anew. As Gilles Deleuze and Félix Guattari (1987) inform, a double articulation is always-already present. For me, this is the both/and nature of induction (i.e., for both science teacher educator and novice science teacher) we must navigate. In this chapter, the two articulations that will be explored are the *sedimentation of a substance* and *folded formations*. Consequently, we begin to see how a doubled and doubling terrain contextualizes meanings and moments. While presented as two separate mechanisms, the sedimentation and folding of the subject are encompassed within two forces. Deleuze and Guattari (1987, p. 40–41) describe the sedimentation of a substance as one that “deducts, from unstable flows” creating substance, while a folded formation “establishes functional, compact, stable structures” or forms. In the sections that follow, I depict two articulations shaping my academic journey: (a) the sedimentation of a lived personal substance, and (b) the formation of folded/ing subject(s) (i.e., novice science teachers). Like geological phenomena, to *think with nature* (Wallace et al. 2018), the social foundations of science teacher induction are laid onto and within the novice science teacher. The following moments interweave present and absent-present multiplicities (Deleuze and Guattari 1987); that is, memories of *a* present which continue to co-construct my academic journey and current research in science teacher education and induction.

20.2 One Articulation: The Sedimentation of Substance

My lived experiences are layered. Like the deposition of sediment, the moments shaping my ideas about science teacher induction are comprised of different consistencies. Moments one, two, three, and four depict the diverse voices, settings, and absent-present-past influencing my becoming a researcher of science teacher induction and educator.

20.2.1 Moment(s) One

"Make sure to demonstrate 'proper methodology.' Justify how you studied yourself. Explain how you ensured trustworthiness and reliability; the other scholars will be interested in this part". While preparing my first research publication in science education, I quickly found myself confronted with the current paradigm of 'good scientific research' even when the edited book I was writing for focused on self-study (Wallace 2016). I was expected to justify the replicable conventions of self-study research, which, as usual, are grounded in positivistic traditions of inquiry that I do not align myself. That is, to reduce inquiry to a strict predictive objective method that ensures a 'clean' procedure and outcome. This was one of the first moments I came to take issue with the assumption that it was possible to claim knowledge of somebody else's experience, and then subsequently proclaim I held the 'right' set of solutions to 'fix' others. I began to wonder how research on science teachers, and the reliance on dominating traditions of 'good science,' also perpetuates distrust in K-12 teachers. For example, when transitioning from being a full-time elementary science teacher to a full-time doctoral student I, like many classroom teachers, was not quick to welcome the advice of educational researchers. Now I see why. Beginning educational researchers are greeted with the same distrust as K-12 classroom teachers. There are particular scientific traditions, with which one ought to conform.

20.2.2 Moment(s) Two

"I don't recognize your citations." In 2016 I presented at the National Assocation of Research on Science Teaching (NARST) annual conference for the first time. The conference caused my stomach to churn many times. Before even attending the first day of sessions, I care-fully read through the conference program to see that the majority of presentations featured the assertion that they had 'figured the science teacher out'. In this moment, I realized that the science teacher had already been deemed a 'knowable entity'; thus, an available *object* to be molded. When presenting my work on science education and science, as a governing forces in my own subjectivity, I was welcomed with blank stares, genuine curiosity (mostly by doctoral students), and assertions that my citations of theorists were 'not familiar' enough (more senior scholars). Additionally, most individuals wanted me to provide concrete results when I often bring only more questions. Throughout the conference, I felt alone. That is, until I ended up sitting next to a British scholar presenting on the new mattering (Barad 2007) in science education. She asked me what my research interests were and then I said the one word that usually makes American scholars' eyes glaze over, subjectivity. For the first time, I felt I had a found somebody who understood me. Our exchange (albeit brief), combined with the scholarly work in international journals where I found most of my scholarly references, caused me to wonder how particular communities of research on science teachers and science teaching actively work to expel the subject altogether.

20.2.3 Moment(s) Three

"*It's like you have a bad critical theory hangover*." When preparing my dissertation proposal, I found myself in a dilemma situated amongst critical ethnography, post-structuralism, and posthuman materialist perspectives. While I realize these intellectual paradigms have their own ontological and epistemological traditions, the notion that scholarly work can and ought to work/live in a particular kind of ideological vacuum causes my stomach to churn and face to scrunch together. As I became more attentive to my emotions during my disseration research, I grew to embrace the opportunity to deterritorialize ontological dichotomies shaping ways educational researchers might un/know the novice science teacher otherwise (Wallace 2017). Don't we all have hangovers of something/everything? What does educational research perpetuate if these theoretical paradigms cannot and should not talk to and within each other? To deterritorialize 'the oughts' of science teacher induction is to re-imagine what might always-already be. Instead of privileging one ontological paradigm over another, I found (and continue to locate) peace within my researcher subjectivity as *both/and*. Rather than intentionally re/territorializing, my dissertation research, and thus inherently myself and implicated participants, I confidently chose to live (and think) in-between the 'oughts' and 'mights' shaping research on science teacher induction. As Jackson and Mazzei (2012) put it, I decided to live and work in the threshold.

20.2.4 Moment(s) Four

"But that's not a novice science teacher". I have been told how I ought to know the novice science teacher. Even as a former K-12 classroom science teacher, I was repeatedly instructed with the caveat, "since this is only your ___ year teaching..." Novice science teachers are often referred to as inexperienced and/or limited to specific stable definitions. For example, scholarly literature has become increasingly over saturated with studies that reduce novice science teacher becoming to their first 3–5 years as a formal teacher. Over the past 30 years, research on teacher induction has been categorized into three tidy definitions: (a) induction as a phase; (b) induction as a process of socialization; and (c) induction as a program of support (Feiman-Nemser 2010). Across each characterization, one theme remains consistent: the understanding of novice science teacher experience is restricted to the first 3–5 years of formal teacher maturation (Bartell 2005). Markers of time (as linear) always-already impose a particular set of assumptions onto the novice science teacher. Within this dominant framing of teacher induction, the novice science teacher will always be positioned as not yet enough. Through this restrictive language, there will always be a bucket needing filled, a mind needing molding, and someone to decide the right ways to do so. The limited and fixed definitions researchers and practitioners regularly work from creates certain possibilities for ways we can and do know the novice science teacher. Each of these definitions begin from a deficit perspective of the novice science teacher subject. Accordingly, as novice science teachers navigate their pre-determined role, they are also constantly 'filled-up' with proper induction supports (i.e., mentors, professional development, science content knowledge). What does this 'filling-up' entail and for what purposes? What enduring implications reside *on and within* the novice science teacher's subjectivity? What implicit (and explicit) violence has taken place? By whom or what? For whom or what purposes? And at what costs? With an eye for ethical relationality, research on science teacher induction and education must think otherwise.

20.3 A Second Articulation: Folded Formations

To work and think anew, science teacher educators and scholars of induction can begin looking to alternative ontological underpinnings. The second articulation of my academic journey specifically attends to Deleuze and Guattari's (1987) process of *facialization*. The concept of facility examines the formations within which subjects become inscribed, literally giving them a "face." While the forthcoming entry points primarily attend to the novice science teacher subject, it also depicts my own being as a 'researcher of science teacher induction and education.'

20.3.1 Alternative Entry Point(s)

By expanding conceptions of science teacher induction to include a process of facialization the ways novice science teacher becomes a body, or face, of signification long before they earn the formal title of "teacher" can be re-engaged. Deleuze and Guattari describe the process in which one gets ascribed a face:

> Facialization operates not by resemblance but by an order of reasons. It is much more unconscious and machinic operation that draws the entire body across the holey surface, and in which the role of the face is not as a model or image, but as an overcoding of all the decoded parts. … *The question then becomes what circumstances trigger the machine* that produces the face and facialization. (1987, p. 170, emphasis original)

The process of facialization depicts how assumptions about what it means to become inducted into the role of 'science teacher' exist within the tense in-between spaces constrained by the *abstract machine*. More specifically,

> It is certain that the signifier does not construct the wall that it needs all by itself; it is certain that subjectivity does not dig its hole all alone. Concrete faces cannot be assumed to come ready-made. They are engendered by an *abstract machine of faciality*, which produces them at the same time as it gives the signifier its white wall and subjectivity its black hole. Thus the black hole/white wall system is, to begin with, not a face but an abstract machine that produces faces according to the changeable combinations of its cogwheels. (Deleuze and Guattari 1987, p. 168, original emphasis)

For novice science teachers the black hole/white wall system *is* the inductive experience. Novice science teacher subjectivities, and also their pedagogical practice(s), get *intentionally* swallowed up, (re)configured, and spit back out. In its current form, the ways researchers assume to know what constitutes science teacher experience and the novice science teacher subject affixes a particular face, which supports the prevailing positivistic underpinnings of science education. We *must* tread lightly or perhaps not at all. Butler (2005) helps to dispel seamless translation of the "I" that supersedes the account one gives of themselves; and in the case of science teacher induction, the account a researcher gives *for* the ways novice science teachers ought to be known and produced. For example,

> After all, under what conditions do some individuals acquire a face, a legible and visible face, and others do not? There is a language that frames the encounter, and embedded in that language is a set of norms concerning what will and will not constitute recognizability. (Butler 2005, p. 30)

Deleuze and Guattari (1987) conceptualize the signification of subjects as a product of facialization within an abstract machine. Societal notions of 'normality' and 'successful,' alongside tidy finite understandings of socialization, inadvertently condition novice science teacher subjectivity. Dominant methods of research on science teacher induction (i.e., surveys, assessment of the success of induction interventions, qualitative thematic analyses), programs of support, and professional development, maintain a banking model of education (Freire 1970) driven by positivism and rationality as the only 'true' and 'right' ontology from which one can be(come) a teacher. Consequently, these ideological underpinnings circulate within the inductive experience. Novice science teachers have been and continue to serve as a site of *intentional* facialization. This process of signification inherently makes its way into the instructional practices novice science teachers enact. This is the abstract machine. For the novice science teacher, "this is an affair not of ideology, but of economy and the organization of power" (Deleuze and Guattari 1987, p. 175). I contend that it is through the abstract machine novice science teachers are expected to successfully and seamlessly navigate, that their teaching ideolog(ies) are manufactured.

It is through the signification of being in-between power/knowledge, the subject, and discourse, that the becoming-novice science teacher is a facialized production. Foucault further describes these machines, or in his terms, apparatuses, as an imposed will to knowledge/truth, "like other systems of exclusion, relies on institutional support: it is both reinforced and accompanied by whole strata of practices such as pedagogy – naturally – the book system, publishing, libraries, such as the learned societies in the past" (1972, p. 219). By beginning to extend perspectives of induction to a novice science teachers' ontological becoming, and signification (or discursive production) the hegemonic structures that shape power/knowledge and thus, the subject, become exposed. Deleuze and Guattari describe,

> That is why we have been addressing just two problems exclusively: the relation of the face to the abstract machine that produces it, and the relation of the face to the assemblages of power that require that social production. The face is a politics. (1987, p. 181)

By deconstructing the inductive experience as a process of facialization researchers of science teacher education and practitioners might engage new ways of critically examining the depths to which "induction" works to constitute novice science teachers and also the institution of schooling. We begin to dismantle the face.

In order to illuminate the (re)inscription of a particular inductive face, researchers of science teacher education must first escape the developmental and militaristic assumptions (e.g., recruitment and training of novice and/or veteran subjects) that neatly sequence teachers being and becoming. The discursive and subjective tension that results from this process is where exciting and innovative possibilities can happen. It is on these plateaus of being that the novice science teacher might be thought of otherwise as *unknowable*. Deleuze and Guattari (1987) are particularly helpful for offering multiple and dynamic ways for researchers of science teacher induction to begin re-conceptualizing ontological possibilities for the novice science teacher subject. Rather than essentializing the subject from traditional Cartesian perspectives as rational and stable, Deleuze and Guattari (1987) open up the possibility of a multiplicitous subject. Jonathan Roffe clarifies Deleuze's concept of multiplicity:

> Deleuze takes the idea that any situation is composed of different multiplicities that form a kind of patchwork or ensemble without becoming a totality or whole. For example, a house is a patchwork or ensemble without becoming a totality or whole. For example, a house is a patchwork of concrete structures and habits. Even though we can list these things, there is finally no way of determining what the essence of a particular house is, because we cannot point to anything outside of the house itself to explain or to sum it up – it is simply a patchwork. This can also be taken as a good description of multiplicities themselves. (2010, p. 181)

Whether as a patchwork piece or nomadic (Roy 2003) exploration, "Once you have used [Deleuzian concepts]... to think in the world, you live differently" (St. Pierre 2004, p. 285). Furthermore, if researchers of science teacher induction and practitioners begin with the assumption that novice science teachers are rhizomatic, multiplicitous, and becoming-teacher (Marble 2012), I wonder (and am excited) to imagine how science teachers might teach differently. Jesse Bazzul and Shakhnoza Kayumova (2015, p. 4) describe "rhizomes and lines of flight escape structures that would seek to capture and reterritorialize them." Given that novice science teachers are expected to successfully navigate a system built to categorize, sort, indoctrinate, and reproduce the rationality of social control and class dominance (Giroux 1980), the work to undo the novice science teacher subject (as we have come 'know' it) becomes a rhizomatic, critical ethical endeavor in research on science teacher induction and education.

By providing a rhizomatic and multiplicitous ontology from which to start reconceptualizing our work with science teacher, Deleuze and Guattari help science teacher educators and researchers ask: How might we engage the preparation and education of science teachers as nomadic and/or following lines of flight?

20.4 Where/Who Now?

Following new lines of flight can be scary. Being a researcher is scary. Educating others is scary. Hence, *educationalways involves risk* (Biesta 2013). To fully embrace moments of hesitation as a generative threshold for thinking science teacher induction and education anew, I have chosen to 'stay scared'. The multiplicitous moment(s) and alternative entry points depict the doubled nature becoming a science teacher educator and researcher of science teacher induction. Evident in Moment(s) One, my academic journey in science education is further complicated by my reluctance to take-up dominant methodological traditions relying on one prevailing conception of good scientific practices in exchange for a messier account found in the post-qualitative paradigm (Lather and St. Pierre 2013). From Foucauldian scientificity (Lather 2010) to educational scientism (Lageman 2000), studies *on* novice science teacher experiences (i.e., the field of science teacher induction and education) continue to be grounded in Enlightenment ideology. Consequently, my academic journey has led me to work differently; that is, by drawing on the abstract realness of Deleuze and Guattari's (1987) work. Their concept of faciality, in particular, complicates the dominant ways research has (and continues) to signify and maintain novice science teachers as a particular kind and properly disciplined subject. At this point, the majority of researchers in science education might assert that my academic journey is doomed from the start; however, I assert that this uneasiness depicts the very issues I strive to debunk (Haslanger 2003) by engaging science teacher induction as also a process of facialization.

References

Barad, K. (2007). *Meeting the universe halfway: Quantum physics and then entanglement of matter and meaning*. Durham: Duke University Press.

Bartell, C. A. (2005). *Cultivating high-quality teaching through induction and mentoring*. Thousand Oaks: Corwin Press.

Bazzul, J., & Kayumova, S. (2015). Toward a social ontology for science education: Introducing Deleuze and Guattari's assemblages. *Educational Philosophy and Theory*, (online first – ahead-of-print), 1–16. DOI:https://doi.org/10.1080/00131857.2015.1013016.

Biesta, G. J. J. (2013). *The beautiful risk of education*. Boulder: Paradigm Publishers.

Butler, J. (2005). *Giving an account of oneself*. New York: Fordham University Press.

Deleuze, G., & Guattari, F. (1987). *A thousand plateaus: Capitalism and schizophrenia*. Minneapolis: University of Minnesota.

Dillard, C. B. (2012). *Learning to (re)member the things we learned to forget: Endarkened feminisms, spirituality, and the sacred nature of research and teaching*. New York: Peter Lang.

Feiman-Nemser, S. (2010). Multiple meanings of new teacher induction. In J. Wang, S. J. Odell, & R. T. Clift (Eds.), *Past, present, and future research on teacher induction: An anthology for researchers, policy makers, and practitioners* (pp. 15–30). Lanham: Rowman & Littlefield Education.

Foucault, M. (1972). *The archeology of knowledge and discourse on language*. New York: Pantheon Books.

Freire, P. (1970). *Pedagogy of the oppressed (30th anniversary edition)*. London: Bloomsbury Academic.

Giroux, H. A. (1980). Critical theory and rationality in citizenship education. *Curriculum Inquiry, 10*(4), 329–366.

Haslanger, S. (2003). Social construction: The "debunking" project. In F. F. Schmitt (Ed.), *Socializing metaphysics: The nature of social reality* (pp. 301–325). Lanham: Rowman & Littlefield Publishers.

Jackson, A. Y., & Mazzei, L. A. (2012). *Thinking with theory in qualitative research: Viewing data across multiple perspectives*. New York: Routledge.

Lageman, E. C. (2000). *An elusive science: The troubling history of education research*. Chicago: The University of Chicago Press.

Lather, P. (2007). *Getting lost: Feminist efforts toward a double(d) science*. Albany: State University of New York Press.

Lather, P. (2010). *Engaging science policy: From the side the messy*. New York: Peter Lang.

Lather, P., & St. Pierre, E. A. (2013). Post-qualitative research. *International Journal of Qualitative Studies in Education, 26*(6), 629–633.

Marble, S. (2012). Becoming-teacher: Encounters with the other in teacher education. *Discourse: Studies in the Cultural Politics of Education, 33*(1), 21–31.

Roffe, J. (2010). Multiplicity. In A. Parr (Ed.), *The Deleuze dictionary revised edition* (pp. 181–182). Edinburgh: Edinburgh University Press.

Roy, K. (2003). *Teachers in nomadic spaces: Deleuze and curriculum*. New York: Peter Lang Publishing.

St. Pierre, E. A. (2004). Deleuzian concepts for education: The subject undone. *Educational Philosophy and Theory, 36*(3), 283–296.

Wallace, M. F. G. (2016). Trash or treasure: Re-conceptualizing my ruins as a tool for re-imagining the nature of science teacher education. In G. A. Buck & V. L. Akerson (Eds.), *Allowing our professional knowledge of pre-service science teacher education to be enhanced by self-study research: Turning a critical eye on our practice* (pp. 341–362). Switzerland: Springer International.

Wallace, M. F. G. (2017). *Deterritorializing dichotomies in teacher induction: A (post) ethnographic study of un/becoming an elementary science teacher* (unpublished doctoral dissertation). Baton Rouge: Louisiana State University.

Wallace, M. F. G. (2018). The paradox of un/making science people: Practicing ethico-political hesitations in science education. *Cultural Studies of Science Education.* https://doi.org/10.1007/s11422-017-9831-3.
Wallace, M. F. G. (in pressA). Becoming-with/in educational research: Minor accounts as care-full inquiry. In K. Strom, T. Mills, & A. Ovens (Eds.), *Decentering the educational researcher in intimate scholarship: Posthuman materialist perspectives*. Bingley: Emerald Publishing Group.
Wallace, M. F. G., Higgins, M., & Bazzul, J., (2018). Thinking with nature: Following minor concepts for ethicopolitical response-ability in science education. Canadian Journal of Science, Mathematics, and Technology Education, 18(3).

Maria F. G. Wallace is an Assistant Professor of Education at Millsaps College in Jackson, Mississippi. Earning a Ph.D. in Curriculum and Instruction with specializations in Curriculum Theory, Science Education, and a graduate minor in Women's & Gender Studies from Louisiana State University, Maria's research and teaching aims to deterritorialize beginning (science) teachers' subjectivities and practice. Drawing on conventional qualitative research methods and post-qualitative modes of inquiry, Maria's research (re)imagines ways beginning science teachers are 'known,' named, and re/produced.

Chapter 21
Pursuing Response-Ability in De/Colonizing Science Education

Marc Higgins

Within school science, Indigenous science (see Cajete 2000) is often either excluded or included in ways that differ from or defer its intended meanings. Differentially, these enact (partial) dialectic negation of Indigenous science by sublating, subsuming, or suturing over it. For Indigenous, diasporic, and other post-colonial students, such school science regularly produces experiences of cultural *assimilation* and *acculturation* rather than *enculturation.* In other words, rather than a harmonious interfacing of cultures (i.e., enculturation), encounters of school science are more likely to house potential for dialectical negation that is either actualized (i.e., assimilation) or remains un-actualized through students' complex and complicated curricular navigation (i.e., acculturation). For these students whose daily lived experiences continue to be negatively impacted by colonial logics (e.g., Eurocentrism), this manifests as a form of epistemic violence. Here, science education and educators have a responsibility.

As an emergent scholar pursing decolonizing science education and aspiring ally of Indigenous peoples the primary orientation that guides my efforts is ethically heeding the call of Indigenous science (e.g., traditional ecological knowledge [TEK], Indigenous ways-of-living-with-Nature [IWLN]). I continue to wrestle with the question: *How is Indigenous science to-come with/in the context of science education*? As Tewa scholar Gregory Cajete (2000) explains, Indigenous science displays and has, since time immemorial, always deployed "ingenuity, creativity, resourcefulness, and ability of people to learn and to teach a harmonious way of existence with Nature" (p. 78). Accordingly, the guiding question I pose is not intended to signal a science yet-to-exist. Rather, *to-come* calls on both the ways in which Indigenous science has not yet (wholly) arrived within the context of science

M. Higgins (✉)
Faculty of Education, Department of Secondary Education, University of Alberta, Edmonton, AB, Canada
e-mail: marc1@ualberta.ca

J. Bazzul, C. Siry (eds.), *Critical Voices in Science Education Research*, Cultural Studies of Science Education 17,
https://doi.org/10.1007/978-3-319-99990-6_21

education, as well as a responsibility of hospitality towards that which is to-come. If science education is to hospitably receive Indigenous science, it must address the ways in which its structures – the assumptions, terms, modes of organization, practices, and beliefs – contribute to exclusion of Indigenous science, as well as inclusion that disciplines, differs from, and defers Indigenous science (to-come). Equally significant is the exploration of how the culture of the discipline of science education be (re)opened and re(con)figured to receive Indigenous science to-come, on its own terms, and in ethical relation? Importantly, when that which is to-come (here, Indigenous science) is never (fully) knowable within and distorted by the current frames of science education, what modes, practices, and enactments of responsibility are available?

Elsewhere (Higgins 2014), I have begun asking similar questions of the relationship between responsibility and the (in)ability to respond within educational research; I refer to this practice as asking questions of response-ability (see also Higgins 2017). This earlier exploration began exposing and troubling the ways in which I was accounting for and being accountable to Indigeneity (e.g., IWLN) from within (naturalized and normalized) (neo-)colonial discourses (e.g., Eurocentrism, whiteness), even though I was actively working against this power differential (see also Higgins et al. 2015). Stated otherwise, as the result of a (neo-)colonial curriculum that is hidden in plain sight, efforts to work against and beyond (neo-)colonial categories, concepts, and structures often come to reify that which is laboured against; decolonizing approaches may come to be de/colonizing (see also Madden and McGregor 2013). In other words, de/colonizing signals the ways in which decolonizing and colonizing are irreducibly co-constitutive. As such, decolonizing cannot be wholly framed in opposition to colonization, at least not within academic and other formal educational spaces given the complexity of their material-discursive structures, even if and when they pursue decolonizing goals. To take seriously de/colonizing is to be hyper-vigilant of the ways in which colonial logics and productions seep into decolonizing efforts. This, as Lyn Carter (2004, 2010) and Ali Sammel (2009) point out, is needed in spaces of science education.

Within this chapter, I continue this exploration with a focus on the relationship between response-ability and my own practice as de/colonizing science educator. This begins with a narrative of a significant personal pedagogical encounter in which the distinction between response-ability and responsibility made itself felt and known. Thinking with the work of Sami scholar Rauna Kuokkanen, this narrative provides a platform to explore practices of epistemic ignorance and their (co-)constitutive relation to science education. It also compels what Kuokkanen (2007) refers to as “the homework of response-ability” required to (re)open the norms of responsiveness towards the possibility of heeding a call of otherness such as Indigenous science from within the structure of science education. Concluding thoughts underscore the promise of deconstruction (rather than destruction) as a theoretical, methodological, and ethical tool to resist the closure of responsibility, through the homework of response-ability, towards hospitably receiving Indigenous science.

21.1 Encountering the Subtle Yet Important Difference Between Response-Ability and Responsibility in My De/Colonizing Science Education Practice

Kuokkanen (2007), whose scholarship centers Western modern educational and institutional responsibility towards Indigenous ways-of-knowing and ways-of-being, defines response-ability as "an ability to respond, to respond to the world beyond oneself, as well as a willingness to recognize its existence" (p. 39). During my first major research project towards decolonizing science education (see Higgins 2014), I was poised to learn a difficult lesson about the distinction between ability and willingness that Kuokkanen presents. My willingness or desire to recognize the otherness that is Indigenous science (in relation to science education) was not sufficient in and of itself. But I am getting ahead of myself here.

At the time of authoring this chapter, the narrative I am about to tell is one that dates almost 10 years. However, it is one that I continue to heed as it continues to bear relevance on how I understand myself in relation to responsibility and the in/ability to respond. During the summer of 2009, I was delivering curriculum that I developed that engaged Indigenous (here, Inuit) and non-Indigenous youth in exploring, constructing, and documenting differential cultural constructions of science (i.e., ways-of-knowing-Nature) through participant-driven videography in their home community of Iqaluit, Nunavut. Through this work, I took up the important call to decolonize science education through what Mi'kmaq scholar Marie Battiste (2013) describes as the "two-prong process" of decolonizing education. It simultaneously and iteratively entails *deconstruction* of (neo-)colonial structures and strategies, and *reconstruction* that centres and takes seriously Indigenous, diasporic, and other post-colonial ways-of-knowing and ways-of-being towards reshaping the place-based processes and priorities of education and educational research. Both prongs are of significance given my positionality as a white, male, fourth generation Euro-settler of Irish and Scottish descent who is working to honour my ever shifting relationships as a science educator working with/in diverse First Nations, Métis, and Inuit communities.

Responding to the first prong (i.e., deconstruction), I engaged in examining and challenging the ways in which Eurocentrism – a pervasive discursive force that (re) centers Western modern(ist) culture, people, places, and histories as the normative standard against which other ways-of-knowing are judged, usually as lesser and deficient (Battiste 2005) – works to maintain the status quo. This was done through working to disrupt the concepts and categories that tend to create, and are utilized to uphold, inequality within science education, as well as the systems under which these inequalities become possible (e.g., "what counts" as science in science education and its entangled apparatus of norms; see Higgins in press). This process largely involved engaging in critical self-reflexive questioning, responding to queries such as:

> How does my Western training in the world of science (i.e., in physics) differentially produce my conceptions of the nature of science, what it is, what it is perceived as, and what it can be? How do I work against the problematic foreclosure of such knowledge in order to maintain pedagogical flexibility? How do I work within and against the implicit Eurocentric notions of validity, empirical worth, and instrumentality that I have received in order to make space for Indigenous knowledges? (Higgins 2014, p. 163)

In engaging in the second prong of decolonizing education (i.e., reconstruction), students were collaboratively involved in creatively juxtaposing Western modern science (WMS) and Inuit Qaujimajatuqangit (i.e., Inuit traditional knowledge) to reveal, (re)structure, and (re)direct the multiple ways that the gaze of dominance is maintained. For example, this gave youth an agentic role in resisting problematic constructions of Indigeneity with respect to ways-of-knowing-Nature (e.g., science). This participant-directed videography took various shapes, notably documentary-style interviews with diversely positioned community members (e.g., traditional knowledge holders, health practitioners, environmental scientists), alongside their own short movies that were a form of digital storytelling.

Despite well-laid plans, in attempting to put to work a decolonizing sensibility and taking up responsibility (both that of pedagogy and educational research), there were nonetheless ways in which I was not able to respond. This subtle but importance differentiation between responsibility and response-ability manifested most noticeably though my self-reflexive work around the youth's engagement in digital storytelling (see Higgins 2014). While my research was originally planned around youth engaging in documentary film, early on the youth made it clear (through scrunched brows, an Inuit way of saying no) that they did not wish to only make movies about their perception of science during what was, for them, a summer camp. Cognizant of the multiple gradients of power across which dissent was being articulated, as well as the ongoing problematic research relationships between research institutions and Indigenous communities (see Battiste 2013), I knew I would have to "let go" of the research project as designed.

While the youth agreed to participate in interviewing community members around ways-of-knowing-Nature, we negotiated that the youth would *primarily* engage in digital storytelling practices as their major project to share with parents and community members by the end of the program. As an emerging decolonizing educator, I recognized the importance of respecting learners' choices. However, as a budding science education researcher, it was difficult to shift away from a focus I had been developing (i.e., exploring cross-cultural ways-of-knowing-Nature). Because youth were spending less time accessing those who 'know' about and with Nature (e.g., traditional knowledge holders, scientists), I had trouble conceptualizing the youth's storytelling practices as enacting ways-of-knowing-Nature. This dissonance was perhaps most heightened when some of the youth explored *Oreo eating Olympics* as a central story topic! In this sense, I was not able to take up responsibility much beyond the ways the ways in which I had conceived of prior the research; there were ways in which I was not able to respond.

Significantly, I had come to the research with the Western modern scientific assumption that *science* is strictly a human, epistemological affair (see Cajete 2000).

As a result, I was blind(ed) to the ways in which some of the digital storytelling practices that youth engaged in (e.g., (re)telling of traditional story of *Muhaha*, the aptly named traditional Inuit monster who chases after children to tickle them to death with his long claws) were not simply stories *about* place but were told *with* place (i.e., having and being had by an Indigenous "sense of place"; see Cajete 2000). In the videos, place makes itself intelligible through the beings that come to (co-)constitute the ecology of relationships that make the eastern arctic a beautiful, yet dangerous place if not respected on its own terms. Their stories 'starred' an ecology of relationships with which Inuit peoples have developed ways-of-knowing-in-being premised on Nature's flux and processes, deeply guided with and through relational ethics, as well as practices of regeneration. The stories were never the students' (and the humans they worked with) alone (despite the frames brought to the viewing); the natural world always makes itself intelligible and participates in the construction of knowledge about itself, whether we acknowledge it or not (see Cajete 1994, 2000).

Elsewhere (Higgins 2014), I stated that the decolonizing curriculum (e.g., border crossing) and pedagogies (e.g., culture broker) available to me worked both within and against a problematic center. As such, curriculum, pedagogy, and pedagogue were exceeded in pedagogical practice by the very coloniality the approach worked against, thus becoming de/colonizing. Here, thinking with Kuokkanen (2007), coloniality overcoded the ability to respond, making me unable to (fully) take up the responsibility of heeding the call of Indigenous science. Specifically, I could not (wholly) respond to the natural world and Indigenous-ways-of-living-with-Nature because I could not recognize its existence (beyond that which made itself intelligible within my frames). I could not responsibly heed the call of Indigenous science because I could not hear the call as such. Again, a willingness to recognize the otherness that is Indigenous science is not sufficient; it is for this reason that Kuokannen suggests that response-ability entails addressing epistemic ignorance.

21.2 Epistemic Ignorance and/in Science Education

Kuokkanen (2010) states, "if knowledge is a prerequisite for responsibility, ignorance presents a serious threat to responsible, response-able behaviour and thinking" (p. 64). Yet, as illuminated by my desire to recognize the existence of Indigenous science from within a Western modern(ist) episteme, working to know Indigenous-way-of-living-with-Nature from such a perspective is also a project that is fraught. As Kuokkanen (2008) suggests, the relation between knowledge and ignorance is not so linear or dichotomous; not knowing is not necessarily an absence of knowledge, but can also be the result of knowledge. Kuokkanen (2008) refers to this knowledge-as-ignorance, as well as traces the discursive forces and flows through which it emerges as epistemic ignorance:

> Epistemic ignorance refers to ways in which academic theories and practices ignore, marginalize and exclude other than dominant Western European epistemic and intellectual traditions. These "other" epistemic and intellectual traditions are foreclosed in the process of producing, reproducing and disseminating knowledge to an extent that generally there is very little recognition and understanding of them. Epistemic ignorance is thus not limited to merely not-knowing or lack of understanding. It also refers to practices and discourses that actively foreclose other than dominant epistemes and refuse to seriously contemplate their existence. Epistemic ignorance is thereby a form of subtle violence. (p. 63)

As Michiel van Eijck and Wolff-Michael Roth (2007) underscore, this is certainly the case in science education regarding the relationship between WMS, TEK, and IWLN. Drawing on Michel Foucault, they explain that the logics of science education can often be characterized as a "regime of truth." Regimes of truth are marked by circular relations: each "truth" is but a differential articulation of the systems of power that produces it, whose articulation in turn (re)produces the systems of power. Such a circular relation can be read in two ways: first, as the capillary circulation of power from one conceptual node to another; and second, signaling a (quasi-)hermetic circle, a (fore)closure of knowledge.

Foreclosure, as postcolonial scholar Gayatri Spivak (1999) utilizes it, signals instantiated pre-emergence of meaning. It indicates the ways in which the language we possess also possesses us. It is when the knowledge shapes how we intake experience, preventing experiences of otherness to be anything more than what can already be known within the already existing, and rigidified, circular relations of closure. In other words, how what we know acts as barrier to engaging with what we do not; a form of closure that is a priori to meaning-making.

The foreclosure resulting from epistemic ignorance is of particular relevance considering the ways in which IWLN and TEK are often only considered science when they fit the criteria of "valid" science (which often happens to be that of WMS). Or, as Kuokkanen (2008) summarizes, the foreclosure resulting from epistemic ignorance makes it such that "Indigenous people 'cannot speak'; that is, when they speak from the framework of their own epistemic conventions, they are not heard or understood by the academy" (p. 60).

On this, Ngāti Kahungunu ki Wairarapa and Ngāi Tahu scholar and science educator Liz McKinley (2007) states, a cartographic relationship between IWLN, TEK, and WMS can be generalized into four categories: (a) where Indigenous science can be explained within WMS; (b) where Indigenous science *could be* explained through WMS, but the explanation has yet to be developed; (c) where there is a link between Indigenous science and WMS's knowledge claims, albeit through different knowledge principles and practices; (d) where WMS cannot accept aspects of Indigenous science (e.g., spirituality, animism). This cartography of relations comes to shape if, as well as when and how, Indigenous science is to be included within school science curriculum. The degree to which 'included' Indigenous science differs from its intended purposes or is deferred through non-inclusion depends highly upon the type of knowledge being brought in, as well as science education's ability to ethically respond to difference (from itself); some forms of Indigenous science are more 'to-come' than others. As the relations of power between IWLN, TEK, and WMS are uneven and unequal, it is too often the case that "those opposing the inclusion [of IK] argue that there is no place for IK unless it has been subsumed into the body of knowledge referred to as WMS, that is, unless it is made the same as WMS, in which the status quo continues" (McKinley 2007, p. 208).

It is for this reason that Kuokkanen (2010) states that "the responsibility toward the other must not emerge from hierarchical relations" (p. 69) as these often come to reproduce the very structures of said hierarchy (e.g., here, the epistemic privileging of WMS over other ways-of-knowing-Nature such as TEK and IWLN). Not only do these hierarchal relations potentially produce foreclosure of (the possibility of) knowing otherwise for those who would wish to uphold the hierarchy, but possibly also for those who are critically within and against it (as illustrated by my story earlier). The reproduction of such hierarchies of relationships (which also remain on-the-move) is not always a conscious choice; even work founded in best intentions to challenge inequitable relations may come to reify problematic structures. Integrating Indigenous science into an educational program that has not come to examine the ways in which it (re)produces and is (re)produced by forms of epistemic ignorance runs the risk of (re)producing similar problematics, albeit differently. For example, this can result in enacting pedagogies or curriculum that work towards "'rescuing' the 'other' or knowing what is best for the 'other'" (Kuokkanen 2010, p. 69). It can also corral Indigenous ways-of-living-with-Nature into a (neo-)colonial space of intelligibility without accounting for or be accountable to the ways in which it differs and exceeds such framing. In turn, the work of responsibility towards the ways in which Indigenous science comes to be othered, as well as to-come, in science education requires more than a desire for the relationship to be otherwise if and when the possibility of ethical relationality is (fore)closed by epistemic ignorance.

There is work to be done: work that addresses not only what we do not know, but also how what we know prevents us from knowing what we do not. This is, following Kuokkanen (2007), the homework of response-ability.

21.3 The Homework of Response-Ability (Towards Indigenous Science) in Science Education

> Doing homework is an ongoing practice that includes learning as much as possible about the area where the academic takes risks. However, familiarizing oneself with areas one knows little about still amounts to hegemonic practice if we do not engage in the "home" part of the homework.... Homework starts from where we are. (Kuokkanen 2007, p. 117)

As stated, earlier, responsibility is often premised upon the possibility of knowing the other(ness) to which we are responding. But, as the discursive formations of science education often come to foreclose the very possibility of (wholly) heeding such a call (through varying degrees of epistemic ignorance), attempting to know about Indigenous science requires that we engage with the "'home' part of the homework" for it to be more than a "hegemonic practice" through which responsibility becomes response-inability through its enactment (even when the individualistic desire is otherwise).

The "home" part of homework can take many meanings: home as cultural, disciplinary, geographical, historical, epistemological, ontological, among others. However, *home*work is always risky as it threatens to rupture who we (think we) are, what we (think we) know, and what we (think we) do. Addressing the ways in which the multiplicity of "homes" in homework are (fore)closed when responding to otherness to-come such as Indigenous science is a project that can be unsettling. In part, this is because it asks critical science educators to examine and sit with the ways in which (their) science education practice continues to uphold problematic practices of subsuming, sublating, and suturing over of Indigenous science. Yet, it must bear risk if we are to (re)open responsiveness and the ability to respond to the (constructed) otherness of Indigenous science which is to-come: "responsibility with an inventive rupture implies, first and foremost, the ability of interrupting the self, of moving beyond the 'I' as the ethical subject" (Kuokkanen 2010, p. 65). Moving beyond the "I" as the ethical subject means considering the ability to respond as being more than individualistic (without excusing the "I" from responsibility). This is a key component of homework. As Kuokkanen (2010) explains, addressing Western modernity as the cultural "home" of science education entails addressing its "worldview of individualism and the notion of the Cartesian subject, [in which] dependency on others is considered a burden" (2010, p. 62).

As science educators, moving beyond the "I" as the ethical subject entails considering the self-in-relation as always already (co-)constituted by vectors of power such as whiteness, Eurocentrism, (neo-)coloniality, modernity, neoliberalism, amidst many others and their respective but irreducibly linked historicities and futurities-to-come, even when working against them. This is all the more important for those who, like me, occupy markers of identity that are privileged by these systems. It is the homework of attending to the ways in which the forces and flows of dominance come to produce the (fore)closure of both self and otherness (making both invisible the normalization of normativity as well as that which lay beyond). Considering the self-in-relation also entails the unheroic work of not assuming that critical pedagogy will be empowering (Ellsworth 1989), but examining the ways in which this approach may always already be disempowering and prevent participants from responding (from 'home'). It is an attempt at a double(d) reversal of the gaze of dominance: a gaze that entails both the literal reversal of studying those who do the studying (i.e., in order to reverse the direction of the gaze), as well as the study of the ways in which those who do the studying study (i.e., in order to reverse the ways in which the gaze is produced and producible).

Yet, while such a double(d) reversal is important, it does not reduce accountability for and towards the other. As mentioned earlier, decolonizing science education must be a movement that creates openings in (neo-)colonial systems and also leverages openings towards making space for honouring Indigenous peoples, places, practices, and priorities. We must still attempt a response within this relation of responsibility, even if response-ability may never be (fully) achieved. Battiste (2005) underscores this (im)possibility using the example of Eurocentrism: "Eurocentrism is not like a prejudice from which informed peoples can elevate themselves" [p. 122]). It is for this reason that there is need to reconfigure the normative processes through which we respond and enact responsibility (within the "home" that is science education):

> What is more, 'starting from here' involves a subtle but radical shift from 'knowing the other' to learning, and more specifically, learning to receive. Rather than assuming the possibility of knowing the other, we need to learn to think in a fundamentally different way. Instead of thinking that 'we must know' or even 'we are entitled to know'—positions that, by retaining the sense of ownership as well as distance, allow very little room for hospitality…—we need to draw a difference, however provisional, between knowing and learning (Kuokkanen 2010, p. 68)

In other words, moving beyond the "I" as the ethical subject entails recognition of the ways in which the Other is always already an irreducible and (co-)constitutive part of the self-in-relation of response-ability. Rather than the individualistic project of knowing the other (which, as mentioned earlier, cannot be disassociated from forms of epistemic ignorance), we are called to learn from the other (something that requires, by definition, relationality). As Kuokkanen (2010) states, not all learning results in knowing the other: response-ability "requires not only patience but acceptance that there will always be gaps, the 'other' can never be fully known" (p. 70). As we learn from the other, we can learn to learn. Stated otherwise, as we attempt to heed the call of Indigenous science (that is not wholly intelligible as such within the epistemes of science education), we must not only listen, but also listen to how we listen (for the ways in which listening prevents us from hearing). It is a subtle and attentive movement that necessarily vacillates between knowing and not-knowing in order to (re)open the norms of responsiveness in order to not only heed the call of Indigenous science but also work towards hospitably receiving this plurality of diverse Indigenous ways-of-living-with-Nature.

21.4 Conclusion: Response-Ability as Moving Within, Against, and Beyond the (Fore)Closure of Epistemic Ignorance or Deconstruction as Learning to Learn

While science education has a responsibility towards TEK and IWLN, it is not always able to enact and uphold this task. As explored within the significant encounter I opened with, the ways in which I became science educator (fore)closed my ability to respond to Indigenous science because I could not heed its call as such. For example, at the time, I could not respond to the ways in which TEK and IWLN are always already more than strictly a human practice that is enacted by the other-than-humans that come to constitute place. My knowledge of what science *is* was knowledge that acted as epistemic ignorance towards what science *could be*; specifically what it always already is and continues to be since time immemorial in the form of TEK and IWLN.

In turn, thinking with this experience, generated the central question explored throughout: How is the irreducible responsibility that science education has towards Indigenous science to be enacted when Indigenous ways-of-knowing-in-being are made unintelligible, undesirable, and/or invisible through science educations' very systems of thought? This chapter coalesces around the (co-)constitutive relation between knowledge and ignorance (e.g., knowledge-as-ignorance) through Kuokkanen's (2008) conceptualization of epistemic ignorance, such that the inability to respond can be framed as rendering unintelligible that which lay beyond epistemic registers, and also inefficacy to account for or be accountable to the ways in which engagement is fraught through the naturalization of said frames. Accordingly, the homework of responsibility through which we labour to transform response-inability into response-ability must entail a (re)opening of the closure through which the other cannot be heard, while simultaneously working to heed the call.

In closing this chapter, it is important to note that addressing the "home" of homework in attempting to move beyond the (fore)closure of knowledge that is epistemic ignorance is not only deeply productive, but also necessary. Addressing the multiplicitous "home" in the homework of response-ability must begin from the ways in which we are shaped by "home" towards its (co-)constitutive exteriority. Kuokkanen (2007) states, as we engage within the very structures that produce epistemic ignorance, this work requires "subtlety and responsibility." Homework that too quickly attempts to evacuate the ways in which we carry "home" elsewhere runs the risk of reproducing the same problems in a new context, albeit slightly differently. Rather, it is important to continue labouring within and against "home" as we attempt the move beyond; to move too quickly to a theory-practice beyond without attending to the "home" of homework runs the risk of as "proceeding in any other way would eventually backfire and merely too tight[ly] reinforce existing structures and discourse [through]… 'irresponsibilizing destruction'" (Kuokkanen 2007, p. xx). In other words, Kuokkanen advocates for a deconstructive and critical inhabitation of these structures rather than an attempt to move beyond through their destruction.

Deconstruction provides a way out of the (fore)closure of knowledge by (re)opening the interiority to its co-constitutive exteriority, and revealing the ways in which the "philosophical category of the center (named Eurocentrism)" (McKinley and Aikenhead 2005, p. 902) operates. It is to attend to porosity between the two in order to displace, disrupt, and decentralize that which was placed with/in (and in turn with/out). As McKinley and her ally Glen Aikenhead state: "deconstruction is the decentralization and decolonization of European thought... Hence, deconstruction is a deconstruction of the concept, the authority, and the assumed primacy of the category of "the West." (2005, p. 902). Deconstruction provides a means of engaging with the interplay of knowing and not-knowing that is inextricably linked to epistemic ignorance, as well as the possibility of placing self and other in relations that (re)open the possibility of learning to learn (as opposed to "knowing the other" within the structures afforded).

Importantly, the possibility of hospitably *receiving* Indigenous science is not only an ethical call. Working to heed the call is ontologically, epistemologically, ecologically, and politically generative: science education stands to learn much from Indigenous ways-of-knowing-in-being and its practices of relational balance, (re)generation, and renewal.

References

Battiste, M. (2005). You can't be the global doctor if you're the colonial disease. In P. Tripp & L. J. Muzzin (Eds.), *Teaching as activism* (pp. 121–133). Montreal: Queen's University Press.

Battiste, M. (2013). *Decolonizing education: Nourishing the learning spirit*. Saskatoon: Purich Publishing.

Cajete, G. (1994). *Look to the mountain: An ecology of indigenous education*. Durango: Kivaki Press.

Cajete, G. (2000). *Native science: Natural laws of interdependence*. Santa Fe: Clear Light Books.

Carter, L. (2004). Thinking differently about cultural diversity: Using postcolonial theory to (re) read science education. *Science Education, 88*(6), 819–836.

Carter, L. (2010). The armchair at the borders: The 'messy' ideas of borders, border zones and epistemological diversity in multicultural science education. *Science Education, 94*, 1–20.

Ellsworth, E. (1989). Why doesn't this feel empowering? Working through the repressive myths of critical pedagogy. *Harvard Educational Review, 59*(3), 297–325.

Higgins, M. (2014). De/colonizing pedagogy and pedagogue: Science education through participatory and reflexive videography. *Canadian Journal of Science, Mathematics, and Technology Education, 14*(2), 154–171.

Higgins, M. (2017). Post-qualitative mo(ve)ments: Concluding remarks on methodological response-abilities and being wounded by thought. *Reconceptualizing Educational Research Methodology, 8*(3), 89–101.

Higgins, M. (in press). Positing an(other) ontology within science education: Towards different practices of ethical accountability within multicultural science education. In K. Scantlebury & C. Milne (Eds.), *Material practice and materiality: Too long ignored in science education*. New York: Springer.

Higgins, M., Madden, B., & Korteweg, L. (2015). Witnessing (the lack of) deconstruction: White teachers' 'perfect stranger' position in urban Indigenous education. *Race Ethnicity and Education, 18*(2), 251–276.

Kuokkanen, R. J. (2007). *Reshaping the university: Responsibility, indigenous epistemes, and the logic of the gift*. Vancouver: UBC Press.

Kuokkanen, R. J. (2008). What is hospitality in the academy? Epistemic ignorance and the (im) possible gift. *Review of Education, Pedagogy, and Cultural Studies, 30*(1), 60–82.

Kuokkanen, R. (2010). The responsibility of the academy: A call for doing homework. *Journal of Curriculum Theorizing, 26*(3), 61–74.

Madden, B., & McGregor, H. E. (2013). Ex(er)cising student voice in pedagogy for decolonizing: Exploring complexities through duoethnography. *Review of Education, Pedagogy, and Cultural Studies, 35*(5), 371–391.

McKinley, E. (2007). Postcolonialism, Indigenous students, and science education. In S. K. Abell & N. G. Lederman (Eds.), *Handbook of research on science education* (pp. 199–226). Mahwah: Lawrence Erlbaum.

McKinley, E., & Aikenhead, G. (2005). Comments on "Thinking differently about cultural diversity: Using postcolonial theory to (re)read science education". *Science Education, 89*(6), 901–906.

Sammel, A. (2009). Turning the focus from 'other' to science education: Exploring the invisibility of whiteness. *Cultural Studies of Science Education, 4*(3), 649–656.

Spivak, G. C. (1999). *A critique of postcolonial reason*. Cambridge, MA: Harvard University Press.

van Eijck, M., & Roth, W. M. (2007). Keeping the local local: Recalibrating the status of science and traditional ecological knowledge (TEK) in education. *Science Education, 91*(6), 926–947.

Marc Higgins is an Assistant Professor in the Department of Secondary Education at the University of Alberta and is affiliated with the Faculty of Education's Aboriginal Teacher Education Program (ATEP). His research labours the methodological space within and between Indigenous, post-structural and post-humanist theories in order to (re)think and practice education which works to ethically respond to contested ways-of-knowing (i.e. epistemology) and ways-of-being (i.e. ontology) such as Indigenous science or ways-of-living-with-Nature.

Chapter 22
Learning About Matter and the Material, Struggling with Entanglement and Staying with the Trouble to Raise Up Feminist Science Education

Kathryn Scantlebury, Anita Hussenius, and Jenny Ivarsson

In this chapter we use metalogue to share and discuss our experiences as feminist science educators using our voices to critique science education, especially its power structures, and each other's thinking while remaining true to an ethic of care that adheres to feminist principles. Kate's feminist pathway began as a secondary school student in Australia and continued through her studies in chemistry and research in science education. For over 20 years she has existed as an 'outsider' in her department, but in the last 7 years, as a visiting research professor at the Centre for Gender Research, Uppsala University (hereafter referred to as the Centre), she has found an academic home and the intellectual space to grow as a feminist researcher.

Anita's feminist awakening occurred through her leadership experiences as head of a chemistry department, which contributed to a major shift in her research interests toward gender and feminist perspectives on science and science education. This includes a problematizing of "science culture", specifically how scientists' conceptions about the discipline and its practices are implicitly and explicitly communicated with students, and its consequences for students' feelings of inclusion/exclusion (Hussénius 2017). Jenny's background is in particle physics and she has spent several years working at CERN. Her move into teaching physics at a Swedish university caused her to reflect on how to make physics more accessible to all learners, especially girls. Karen Barad's groundbreaking book *Meeting the Universe Halfway* (2007) inspired Jenny to explore the influence of new materialism on

K. Scantlebury (✉)
University of Delaware, Newark, DE, USA
e-mail: kscantle@udel.edu

A. Hussenius
Uppsala University, Uppsala, Sweden

J. Ivarsson
Department of Natural Sciences, Bessemerskolan, Sandviken, Sweden

J. Bazzul, C. Siry (eds.), *Critical Voices in Science Education Research*, Cultural Studies of Science Education 17,
https://doi.org/10.1007/978-3-319-99990-6_22

learning in physics and science education. Though we have taken different pathways, over the decades we have stayed with the trouble through our engagements in science education and feminist research.

There are different feminist theoretical frameworks (e.g. radical, liberal, Black, Chicana) and what they have in common is critiquing how societal power structures oppress females and other marginalized groups. Through our experiences and involvement with feminist research, we are engaged with examining gender issues. While, gender is always a component of research, our research is not only about gender. Mari Matsuda (1991) raised these questions

> When I see something that looks racist, I ask, "Where is the patriarchy in this?" When I see something that looks sexist, I ask, "Where is the heterosexism in this"? When I see something that looks homophobic, I ask, "Where is the class interest in this? (Matsuda 1991, p. 1189)

Matsuda (1991) captures the multifaceted complexities that may arise, when the discrimination target needing visibility and attention, always could be something else. Today, the intersections between several social categories are central to contemporary feminist research, which thereby contribute to new insights that had not been possible to achieve by addressing one category at a time (Hill Collins and Bilge 2016). In this chapter, a feminist stance provides common language for our discussion of engaging with material feminism, the entanglement we have experienced in connecting with this post humanist theory and how it has enabled us To "Stay With The Trouble" (Haraway 2010).

22.1 Movement: Space: Language: Discourse: Culture: Matter

Kate: As a feminist, I examine whether science education is moving forward on issues related to girls and women and at the Centre I found an academic home/department/place that I had never experienced before. My metaphor for the Centre is a 'feminist paradise', a space where one moves beyond explaining that gender is a social category, or that there is no 'one' feminism to engage in challenging conversations to critique one's ideas. As an intellectually diverse, rich and interesting space, the Centre's researchers come from humanities, social and natural sciences, and medicine with a range of research interests. They bring to the Center a strong theoretical grounding in the discipline commensurate with a commitment and passion to engage in gender research with an open mind for various theoretical frameworks (for example feminist, masculinist, queer, human-animal, post-modern/structuralist/humanist, technosciences, and new material feminism), methodological approaches and cultural views. The scholars are supportive of one's ideas, producing an intellectually safe space to think aloud, to seek critique, and to share perspectives and direction on one's research.

Anita: Also I appreciate the interdisciplinary and open-minded atmosphere at the Centre, although for me it is not mainly a place to "rest" as a feminist. Rather it is an environment challenging disciplinary boundaries and as such a place to rest from narrow-minded scientific put-downs on what "real" research is and how it should be conducted. Unlike Kate, who regularly visits the Centre and then returns to her chemistry department, I have permanently moved from a chemistry department to a natural sciences department (biology, chemistry, physics, mathematics and computer science) and then to this environment with a majority of humanities and social sciences scholars. It has definitely broadened my mind and introduced new methodologies and theoretical frameworks in my research. Even more importantly, it has opened up an increasing awareness of culture characteristics that differ among disciplinary domains as well as among separate disciplines within the same domain. This awareness goes beyond things that are visible, audible and perceptible, what Edgar Schein (2006) called artefacts and is the more easily recognizable aspects of a culture. Instead it concerns shared norms, values and core presumptions, where the latter are so deeply rooted that those belonging to the culture take them for granted (Schein 2006).

Through Cathrine Hasse, I learned about the method of culture contrast (2015). We use this method in a research project, *In the borderland between academic disciplines and school science – Science faculty as teacher educators,* to examine how science epistemology, content, and practice is reproduced and transformed when scientists engage in teacher education. We collect empirical data by shadowing science faculty in their daily activities and conducting individual and cross-disciplinary focus group interviews. The culture contrast method is theoretically underpinned by a conceptualization of practices as cultures, with implicit and explicit rules, and different values and underlying assumptions. With this method it is possible to see the dominating or ignored cultural values and norms. The different actors that are part of a culture are understood as "carriers" of the culture's underlying ideas, which in different ways are manifested in their actions and through these actions possible to get sight of (Hasse 2015).

***Kate*:** There is limited science education research using feminist and queer theories. This chapter focuses on our struggle to engage with post-human theories and push our thinking into ways that science education research could use material feminism. Barad's simple sentences introducing her article on post-humanist performativity encapsulated these concerns.

> Language has been granted too much power….. Language matters. Discourse matters. Culture matters. There is an important sense in which the only thing that does not seem to matter anymore is matter. (Barad 2003, p. 801)

Anita: I know that these sentences are often cited and capture in a condensed way a critique of the strong influence the linguistic turn has had in the humanities and social sciences. But is this criticism accordingly directed towards natural sciences and science education? For me it has been, and still is, hard to understand the way in which the language has gained more power at the expense of matter within science, including science education. As a former chemist and chemistry educator, everything concerned matter – my practices in the laboratory and the content I taught. Of course language was important, but mainly the chemistry language: the concepts and the ability to communicate through chemical formulas. Upon moving to the Centre I widened and shifted that focus to include language, discourse and culture. I am not troubled by the emphasis on discourse – for me it has opened my eyes and made me aware of aspects I did not notice before.

Kate: During a sabbatical leave at the Centre, I was updating a 2007 handbook chapter focused on the research in gender and science education (Scantlebury 2014). I thought material feminism had something to offer to science education. But what? In the late spring, our group went to Skagen for a writing retreat. Over the centuries, artists have travelled to Skagen to paint the landscape, and revere the beauty of where wind and water intermingle to produce a salty spray when water from the North Sea becomes entangled with Denmark's straits. Walking along beaches and over sand dunes or into the surrounding forest placed one 'of the world' and the experiences began to help me understand Barad's sentences and the implications of accepting matter's agency.

I had struggled to understand Barad's concepts of material-discursive practices, agential realism, phenomena, apparatus, spacetimemattering, and onto-logical-epistemology (Barad 2007). Fortunately, several scholars at the Centre used Barad's theory and were happy to discuss these concepts. I continued to struggle with the ideas but decided the way forward was presenting my initial thoughts about how science education could benefit from material feminism at an informal presentation to my colleagues involved with the writing group at Skagen. My initial argument did not convince them, they raised questions and posed challenges, some I could answer, many I could not. After we left Skagen, I prepared a seminar for the Center's spring series on "how to make matter matter to science education".

22.2 Struggling to Understand Material Feminism

Anita: I was one of those not convinced. In my initial reading of *Meeting the Universe Halfway* (Barad 2007), I interpreted her critique of the linguistic turn as a criticism that did not concern natural sciences, as most scientists do not know what the "linguistic turn" is. That Barad, as a physicist, wanted to challenge the non-scientists who had become captivated by the linguistic-turn rhetoric and discourse. Secondly I interpreted that her theory included a criticism of the neglecting of "discourse and culture" within natural science. In that, I was thrilled and challenged through discussions with a particle physicist, Jenny. What Barad provides, leaning on Niels Bohr, is an empirical basis by reintroducing the importance of matter, but with the phenomenon as the foundation instead of atoms, molecules and natural forces (Ivarsson 2016). In addition to our discussions, Jenny gave a seminar at the Centre for Gender Research. One thing led to another and half a year later both Kate and Jenny were involved with an international graduate course on gender and science education, *Intersectional perspectives on science and science education* and our discussions continued (Fig. 22.1).

Jenny: Classical physics is based on essentialism, the assumption that the world is composed of independent objects with intrinsic properties. It is also based on representationalism, meaning that the observer studies the object from a distance, without affecting or being affected by it and then produces a representation, having

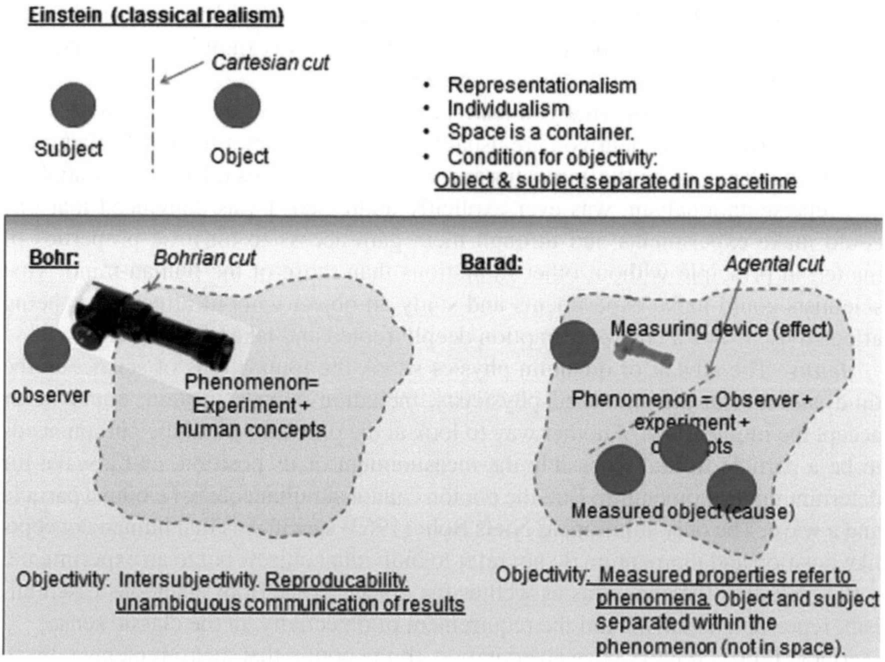

Fig. 22.1 A slide from Jenny's seminar

no direct access to the object itself. The requirement of objectivity is irrefutable in science and in classical physics, which means that object and subject must be separated in space-time. Thus in a classic measurement, there is a clear distinction between the observer and the observed.

If you observe subatomic particles you enter the world of quantum physics. It will become clearer what an act of measurement actually is and it will not conform to the classic notion of objectivity. Quantum effects are admittedly difficult, or practically impossible to detect on a macroscopic level, but in theory, quantum mechanics applies at all levels. From an ontologist's perspective, it does not matter if the object being measured is a macroscopic object or an elementary particle. So, let us look at a simple position measurement. Position measurement means that at least one light-particle, a photon, must strike the object and then be recorded. The photon's position must be registered by a rigid photodetector. If the photodetector is not fixed, the image becomes blurred and the position will not be exactly determined. However, from a classical perspective the photon disturbs the object by means of a momentum transfer. The momentum of a particle is related to its motion. Whenever momentum is transferred from one particle to another, the speed and direction of motion is changed. The exact position of the object cannot be known unless this is compensated for. But in order to determine the momentum transferred, the momentum of the photon must be recorded. Such a measurement requires a photodetector, sensible to motion – not a fixed one. The requirement of the equipment to be able to measure the photon's momentum cannot meet with the requirement for measuring its position. In other words, the photon cannot simultaneously be part of the object and part of the instrument. There is no way to measure both position and momentum of a particle at the same time. Position and momentum are examples of complementary concepts that are intertwined, entangled.

Anita: That is part of what you learn at undergraduate level both in physics and chemistry. But as a student you are usually left there. I cannot remember that these conflicting demands on the apparatus and what consequences it has for essentialism and representationalism, was ever explicitly addressed. I was convinced that one could make experiments and through them gain access to intrinsic properties of matter, in principle without other limitations than those of the human mind. That scientists could make experiments and study an object without affecting or being affected by it, was a core presumption deeply rooted and taken for granted.

Jenny: The advent of quantum physics shook the foundations of science at the time and several distinguished physicists, including Albert Einstein, could never accept the implications. Another way to look at the problem is that the photon needs to be a particle to make possible the measurement of its position, and a wave for determining its momentum. But, the photon cannot simultaneously be both a particle and a wave. The only solution, as Niels Bohr (1963) saw it, was that human concepts like position and momentum do not refer to individual objects but to an experimental setup including the apparatus as well as the object. Thus, Bohr abandons essentialism, representationalism and the requirement of objectivity, in the classic sense.

Barad (2007) reintroduces objectivity with the notion that measurements refer to phenomena, which means that they are reproducible and communicable and put permanent tracks – the photon "becomes" either a particle or a wave, depending on the experimental setup. At a first glance this may seem paradoxical, but the paradox arises only in a representational paradigm that assumes an abstraction of a photon, existing before the act of measurement. If we let go of our 'cultural backpack' and try to focus on what is really going on, it is just measurements and the only things observed are phenomena. The idea of a pre-existing photon is a model that was never confirmed by observation. So, if phenomena are all we observe, then why not consider phenomena to be the stuff the world is made of? You might still object that we don't know what is going on when matter is left alone, if no one is measuring on it. Such an argument relies on a distinction between human and the rest of nature. Denying such a distinction leads to the understanding that our interaction with particles is not different from when particles interact with each other. In this respect, post-humanism actually renders human a higher status and a direct access to the ontology of the world.

If science would embrace post-humanism's view, there would be no need to learn how to compensate for the influence of the observer and more effort could instead be spent on investigating and understanding the role of the observer as an operative part of the phenomena. The scientist would be an integrated element in science education and science would, I believe, appear more accessible to non-scientists.

Anita: In science education we use and talk about models, stressing that it is a model. But in my (former) view a model was something that could be improved, that over time would develop and eventually reach a true description of some intrinsic property of matter/nature. For me the Baradian view of phenomena, opened my mind to the possible usefulness of feminist materialism in science education. Through my discussions with Jenny, I became gradually more interested in Kate's arguments and wonderings about how science education could benefit from material feminism. Actually, writing this narrative piece, forcing myself to critically listen to my own inner voice, has been an awareness opening process. I have asked myself why I initially reacted with skepticism instead of a non-judging open-minded interest, where the latter is something characteristic of most scholars at the Centre for Gender Research. My conclusion is that I found it a bit strange that I came in contact with a physicist's theoretical construction in a research environment where a majority of the scholars are humanists and social scientists. Before Kate raised it, others had embraced this agential realism and "materiality turn", making it a topic for discussions and seminars. I am embarrassed to admit that these scholars probably did not have sufficient legitimacy in my eyes, regarding Barad's work. But more important, I did not understand their interpretations and explanations, the way they talked about her theory; I did not understand their language. This inability to understand was the base that shaped my resistance, which Kate met when she wanted to share her growing interest and wonderings about material feminism. In retrospect I can see the crucial impact of the physicist Jenny, to guide me into Barad's text. I needed someone that catalyzed the dismantlement of my initial wall of resistance, someone who spoke about matter and material feminism with a language familiar to me from basic physics. I realize that my science cultural backpack was and still is a blinder that I need to scrutinize. For my own part, I also think it is important not to rule out the power of language. Maybe, if we think of Barad's critique of the linguistic turn "Language has been granted too much power" (Barad 2003, p. 801). But we must not forget the power that language does have. Language is a part of the culture; language is a part of the discourse. So, what if we consider the entanglement between matter and language?

Jenny: I recognize myself in Anita's resistance. The suspicion among science scholars against any research in discourse probably depends on the strong opposition between spoken language and matter. Should matter itself "merely" be a concept? To me the crucial point was when I realized that an experimental setup is a kind of discourse too. The instrument is the missing link between matter and discourse as well as between social sciences and natural sciences. Bohr had an understanding of the status of measurements. For example, there is no "position". The only thing there is, is a position measurement. Bohr did not express those insights within the science discourse as stringently as Barad did, supported by the post humanism theoretical construct. Whether Bohr actually thought the way Barad argues is irrelevant. What is interesting is the ontology Barad produces through her interpretation.

22.3 Staying with the Trouble: Challenging Science and Science Education's Meta-Narratives

What is to be learned in our narratives? There are few feminist voices in science education research, and to have critical colleagues with this stance supporting us to stay with the trouble is helpful. It is an on-going and daily challenge to live, work and enact feminism (Ahmed 2017). Academe is well documented as a structure that reinforces masculine hegemony, and the culture within science departments is often hostile to theories that challenge the assumption that these disciplines are rational, logical and value free. By engaging gender scholars with science education, we establish a research space to focus on how material feminism can make matter matter in science education, offering an enriched understanding of matter.

Kate: For me, the next step in this narrative was the opportunity to write a 'wonderings' paper as my 'ticket' to enter into a Cultural Studies of Science Education (CSSE) workshop at the University of Luxembourg. I wrote on material feminism and how matter should come to matter. During the workshop, I talked with Cath Milne about my ideas and struggles with understanding "Baradian" concepts. Cath suggested we propose symposia to NARST and the American Educational Research Association (AERA) on the role of matter and materiality in science education. Her interests focused on the role of instruments, she knew of other scholars who were also engaging with the material. Cath led the preparation of the symposia proposals, bought her own copy of *Meeting the Universe Halfway* and an outcome of the ongoing conversations with Cath, and the other scholars involved in the symposia, was an edited book. The supportive networks at the Centre, and also through the scholars involved with CSSE, have provided multiple opportunities to discuss my understanding of how to use material feminism in science education research.

Anita: My moving from 'pure' natural science departments to an environment hosting mainly scholars from humanities and social sciences made me a carrier of other practices and perspectives that differed from the majority of those belonging to this research milieu. I ended up in an interstitial space that provided me an opportunity not only to experience but also challenge this, for me at the time, new culture. But maybe even more important for my individual scholarly development, it offered me a possibility to acquire new practices and cultural awareness. I do not mean that it is consequently necessary to move, to get sight of one's own cultural backpack. Although this has been my journey, I do not think that this is a necessary condition for gaining awareness of those core presumptions that are taken for granted. However, norms, values and core presumptions shared by members within a discourse, within a culture, need to be made explicit in one way or another; this is especially important if they are part of power structures that ostracize potential participants depending on gender, ethnicity, sexuality, functionality or any other basis for categorization.

Working in interdisciplinary collaborations has raised my cultural awareness. Kate's friendly, feminist and persistent voice has not left me unmoved but pushed my thoughts in new directions, that otherwise had been closed. Barad's theoretical contribution has broadened my understanding of matter and phenomena; it has added a discourse dimension to my view of basic chemistry research. Or as Jenny puts it, an experimental setup is also a kind of discourse; the instrument/apparatus the missing link. Research into science discourse, in laboratory settings, has previously failed to engage with, examine, and understand the influence of instruments. The power of instruments is rarely examined, yet the instruments are an important aspect of the discourse in a research group. This discourse is also changed when new instruments are added to the setting (Pettersson 2011). Moreover, the instrument, and the discourse produced can differentially reveal matter, producing unique and varying phenomena. This understanding of the laboratory discourse and the power of the instrument presents a challenge of science and science education.

Kate: While post humanities research seeks to decenter focus on the human, much of the current research and writing does not succeed in this regard. Science education research has not engaged with the post human and in particular with how these theories could influence a culture that remains masculine, white and Western centric. The challenge is to "stay with the trouble" and identify ways to use material feminist theories as a framework for re-directing science education research. ***Jenny:*** Many theories of identity formation presuppose that humans have a special position in nature. There is a human essence, according to which only humans have agency and ethics. As a consequence many standpoints will be highly dependent on what should be classified as human (When does the fetus become a human? Can a robot be human?). Barad (2007) argues that there is no distinct boundary between human and non-human. This applies to all boundaries (human-animal, life-death, object-subject). Boundaries are defined within a phenomenon. As such they are real, well specified and not arbitrary, but they are not given from above and not fixed once and for all.

22.4 Raising One's Voice

It is important for feminists to raise our voices whenever systems of oppression come to the fore, especially within educational settings. There are issues that impact the lives and opportunities for girls and women and delay the progress toward an equitable society. Disenfranchising power structures will continuously need to be identified, challenged and dismantled. The multifaceted complex of problems and the new insights on these problems gained by intersectionality studies have added a new dimension of knowledge compared to more narrowly defined studies. This analytical perspective brings to the fore the diversity and complexity of female's social context. The various axes of social division can reinforce disenfranchisement of females from science and science education. Yet as white, middle class, educated women we are aware of the privilege and position. But this privilege should not be taken as a motif to silence our voices, just like the Matsuda quote, in the beginning of this chapter, should not be interpreted in a way that silence voices. From our point of view the main problem is the lack of feminist research in the wider research community as a whole and in science education specifically. Every voice addressing this deficit is an important voice.

In order to reach acceptance in science culture it is helpful to illustrate how feminism can contribute to the progress of science itself. The relevance of feminist perspectives is obvious in connection to biological concepts that are associated with ethology, interpretation of behavior, species reproduction etc. By challenging what is natural, scientists have increased opportunities to study biological phenomena. Traditionally, observations of nature were interpreted through a gendered stereotypic lens, which focused on studying males. This myopic view resulted in scientists being oblivious to other phenomena such as females establishing territories or the implications of female coloration (McLennan 2000). The introduction of feminist perspectives has opened up new fields of research and has lead to a higher acceptance of feminism in biology (Ah-King 2013).

We see a potential for an analogue progress in other areas. Barad's contribution to the philosophical interpretation of quantum physics would be an example from the field of physics. Unfortunately, those insights are difficult to access, so a catalyzer from other areas in physics, chemistry or related disciplines is needed, where a feministic stance would put traditional scientific issues in a different light and lead to progress in the understanding of particular phenomena. Could the Baradian perspective of the instrument be a part of that catalyzer?

Virginia Wolff articulated the importance of an intellectual and physical space for women to write, think and agitate in her essay *A room of one's own* (Wolff 1991). And Adrienne Rich's feminist essays remind us to claim our education and the responsibility to one's self and others from that education (Rich 1979). We will continue to motivate researchers to engage with gender perspectives by taking the responsibility to raise our voices, offer critiques, identify opportunities for collaboration and exploration into new research areas where gender and feminist perspectives are included and taken seriously.

References

Ah-King, M. (Ed.). (2013). *Challenging popular myths of sex, gender and biology*. Crossroads: Center for Gender Research, Uppsala University, Springer Publishing. https://doi.org/10.1007/978-3-319-01979-6.

Ahmed, S. (2017). *Living a feminist life*. Durham: Duke University Press.

Barad, K. (2003). Posthumanist performativity: toward an understanding of how matter comes to matter. *Signs, 28*(3), 801–831. https://doi.org/10.1086/345321.

Barad, K. (2007). *Meeting the universe halfway: Quantum physics and the entanglement of matter and meaning*. Durham: Duke University Press. https://doi.org/10.1215/9780822388128.

Bohr, N. (1963). *The philosophical writings of Niels Bohr. Vol. 2. Essays, 1933–1957, On atomic physics and human knowledge*. Woodbridge: Ox Bow Press.

Haraway, D. (2010). When species meet: Staying with the trouble. *Environment and Planning D, 28*(1), 53–55. https://doi.org/10.1068/d2706wsh.

Hasse, C. (2015). *An anthropology of learning: On nested frictions in cultural ecologies*. Dordrecht: Springer. https://doi.org/10.1007/978-94-017-9606-4 10.1007/978-94-017-9606-4.

Hill Collins, P., & Bilge, S. (2016). *Intersectionality*. Malden: Polity Press.

Hussénius, A. (2017). Forming bonds – breaking bonds. In L. Bryan & K. Tobin (Eds.), *13 questions for science education*. New York: Peter Lang Publishing.

Ivarsson, J. (2016). Agentisk realism, en genusvetenskaplig formulering av kvantfysikens lagar. *TGV, 37*(3), 96–113.

Matsuda, M. (1991). Beside my sister, facing the enemy: Legal theory out of coalition. *Stanford Law Review, 43*, 1183–1192. https://doi.org/10.2307/1229035.

McLennan, D. A. (2000). The macroevolutionary diversification of female and male components of the stickleback breeding system. *Behaviour, 137*(7), 1029–1045. https://doi.org/10.1163/156853900502411.

Pettersson, H. (2011). Making masculinity in plasma physics: Machines, labour and experiments. *Science Studies, 1*, 47–65.

Rich, A. (1979). *On lies, secrets and silence: Selected prose, 1966–1978*. New York: Norton.
Scantlebury, K. (2014). Gender matters: Building on the past, recognizing the present, and looking towards the future. In N. Lederman & S. Abell (Eds.), *Handbook of research on science education* (Vol. II, pp. 187–203). New York: Routledge.
Schein, E. (2006). *Organizational culture and leadership*. San Francisco: Wiley.
Wolff, V. (1991). *A room of one's own*. New York: Harcourt Brace Janovich.

Kathryn Scantlebury is a professor in the Department of Chemistry and Biochemistry at the University of Delaware, Director of Secondary Education in the College of Arts and Sciences. Her research interests focus on gender issues in various aspects of science education, including urban education, preservice teacher education, teachers' professional development, and academic career paths in academe. Scantlebury is a guest researcher at the Centre for Gender Research at Uppsala University, Co-Editor in Chief for the journal Gender and Education and co- editor of two book series for Sense Publishers. Her email is kscantle@udel.edu.

Anita Hussénius is director of the Centre for Gender Research at Uppsala University and associate professor in Organic Chemistry at the University of Gävle, Sweden. Her main research interest is about gender and feminist perspectives on science and science education. More specifically, her research focuses on issues connected to gender awareness in science and in science teaching. This includes a problematizing of the "science culture", that is what/how conceptions about the discipline and its practices are implicitly and explicitly communicated in the meeting with students, and its consequences for feelings of inclusion/exclusion. Her email is anita.hussenius@gender.uu.se.

Jenny Ivarsson holds a PhD in high energy physics. She has spent several years at the European research center for high energy physics, CERN. Her current research interest is in science education and gender perspectives. Ivarsson has also a secondary science teacher degree. She has six years of experience as a lecturer at Gävle University and is currently teaching in upper secondary school, Bessemerskolan, Sandviken. Her email is jenny.ivarsson@edu.sandviken.se.

Chapter 23
Pushing the Political, Social and Disciplinary Boundaries of Science Education: Science Education as a Site for Resistance and Transformation

Carolina Castano Rodriguez

Coming from South America, revolutionary and emancipatory thinkers such as Che Guevara and Paulo Freire have shaped my identity as an academic. Their ideas have provided me with an ample lens towards resilience, inclusive education, diversity, transformative learning and emancipatory education. However, the academic professional practice and structures I have to participate in often confront my commitment towards social and ecological justice. The diverse issues I face to continue my commitment towards social justice makes me questions whether the work of academics validate the limiting views of what is of value for western countries.

Moreover, despite research in science education that calls for educating more critical and informed citizens (Hodson 2003), there are limited opportunities, I find, for science educators to generate change, transform communities and empower those that have been disempowered by society. I argue that science education needs to be founded in the emancipatory theories of resistance to radicalize its aims and practices as a way to move forward our social justice agenda. This needs to start by furthering our understanding of "the relationship between the system that we oppose and the way in which it shapes our opposing acts" (Jaramillo and Carreon 2014, p. 407).

In this chapter, I use my personal narrative to envision a way forward towards a critical science education that takes up the role of transforming lives and communities by critically analysing how pedagogies of resistance, an ethic of care, and radical peace education can help educators rethink science education as a site for resistance and transformation. The use of personal narratives in education scholarship originated, in part, as a form of resistance to perpetuating discourses and genres which are "hyper-theoretical" and difficult to access (Reed-Danahay 1997). Special consideration is given to the links between theory and practice, and the intersections of collective and individual experience, struggle and construction of identity(ies) (Reed-Danahay 1997).

C. Castano Rodriguez (✉)
Australian Catholic University, Melbourne, Australia
e-mail: Carolina.CastanoRodriguez@acu.edu.au

J. Bazzul, C. Siry (eds.), *Critical Voices in Science Education Research*, Cultural Studies of Science Education 17,
https://doi.org/10.1007/978-3-319-99990-6_23

I conclude that to advance the social justice agenda we need to transverse the boundaries of culture, discipline, political and social institutions, and radicalize our discourses and practice in science education toward peace and care for all.

23.1 Becoming an Activist Academic in the Era of Corporate Education

I grew up in Colombia in the 1980's and 1990's when violent attacks, killing, bombs and kidnapping was common. Colombia still face many issues with violence and in 2016 there were still 7.9% Colombians living in extreme poverty (Departamento Administrativo Nacional de Estadisica – DANE 2016). My cultural background and experiences have shaped my interest in helping others and empowering marginalized and underrepresented communities. What I experienced growing up in Colombia has also provided me with skills to engage in practices of resistance. Following Freire's (1973) emancipatory theory in which education is a resource to mobilize minority and poor communities, I strongly believe that the work of academics could contribute to transforming lives. Academics should be activists. Universities should generate and protect spaces for controversy, critical thinking and debates to push the boundaries of societies, to contribute to create more just societies.

Transformation is a word I have heard often in educative discourses: transforming children, transforming practices, transforming society, etc. However, transformation is more than just a popular concept. There is an entire theory of transformative learning which has contributing to my understanding of how I perceive education and how experiences I had shape my views and challenges. Transformative learning theory explains how triggering and confronting events are fundamental for transformation of thought and actions to occur (Mezirow 1991). However, in the era of corporate education (Thomas 2013) there seems to be reduced spaces for academics to promote transformation. Nationalization, standardization and censorship of what is allowed to be presented and discussed in classes is increasingly becoming a focus of the educational policy and institutions. This reality was identified more than three decades ago by critical theorist Michel Foucault (1984). Surveillance strategies, such as testing and certification, have prevailed and proliferated. Surveillance has extended beyond schools to universities, with freedom, divergent thinking and criticality compromised:

> It is the individual as he[she] may be described, judged, measured, compared with others, in his[her] very individuality; and it is also the individual who has to be trained or corrected, classified, normalized, excluded, etc. (Foucault 1984, p. 203)

Today, this means schools by and large push for an inexpensive and compliant ideal worker. For those academics interested and guided by a commitment to contribute to a more socially just society it is important to ask ourselves: how could we provide 'someone else's shoes' for our students to try while facing the challenges posed by the current trends of academia? Is an activist academic an utopic dream?

Following a Social Resistance Model (Factor et al. 2011) which aims to explain behaviours of non-dominant groups integrating structural and agency factors, I position myself as a member of a non-dominant group in the international academic context. I identify myself as a South American, to use the geographical location, rejecting the notion of "Latino(a)" as limiting and non-reflective of the diversity of cultures that represent South America. I also oppose the trend to classify who we are and what we do as academics in discrete and limiting groups. As explained by Social Resistance Model, actions, activities and behaviours of non-dominant groups which do not follow the expectations of dominant groups are a form of active resistance. Rather than viewing individuals from non-dominant groups as passive agents, whose actions are consequences of power relationships, Social Resistance Model explains "unhealthy" choices and high-risk behaviours of non-dominant groups as forms of resistance; as active agents. Despite the focus of Social Resistance Model on "risk" and "unhealthy" behaviours, such as stealing, it could also serve to explain how choices which are not evidently unhealthy are also active forms of resistance rather than passive consequences of power relationships. Following this, I reject the notion of, and do not identify myself with, "Latinos" from poor countries who are underrepresented in the international academic context and who need "help" and further opportunities to increase representation. I position myself as an active agent from diverse non-dominant groups (female, from South America, with mix heritage of Indigenous, African and Spanish cultures) whose actions and decisions contribute to resist hegemonic trends.

In what follows I analyse how confluence of my socio-cultural background of a South American, the opportunities and challenges of working in Australia, and the tradition of objectivism of science works to create sites of resistance and transformation. I focus on my socio-cultural background as leverage for creating spaces of resistance and agency and present diverse ways in which I engage with diverse forms of everyday Social Resistance action and the tensions I also have to navigate. As described by Patricia Ewick and Susan Silbey (2003) everyday resistance are actions that individuals from non-dominant groups adopt to adapt to power and dominant structures while protecting their own interests and identity. They react to dominant structures and power relationships in ways that reinforce the separation from the dominant group and their collective identity (Ogbu 2004).

23.2 Contrasting Experiences in Colombia?

Colombia has endured more than five decades of internal conflicts with one of the longest lasting guerrillas, FARC, with whom the government is currently negotiating a peace process. It currently sits as only second to Syria in the number of victims of internal displacement (International Displacement Monitoring centre – IDMC 2015). By contrast, Colombia also benefits from highly prestigious public universities and is recognized by the positive attitude to life and happiness of its citizens (Worldwide independent network of market research – WIN and Gallup International 2015).

This dichotomy between challenges and positive qualities of Colombia have contributed significantly to my interest for social justice and increased my understanding of adversity and resilience. I view education as transformational and search for engaging in activities that provide opportunities for contributing towards social justice. It is a challenge to hold this view while at the same time I am part of the international education system that focuses so much on measurable global outcomes rather than the local needs. However, despite these challenges which I go more in depth in the next session, I have been able to find opportunities to engage in social justice projects, by going beyond the expectations and restrictions of academia. For instance, during 2016 I was invited by the Universidad Nacional de Colombia, a public university, to design and offer a course in science education for their International School. I was trusted to create a course on science education that could be relevant for the current social situation of Colombia. That was the main purpose: relevance. I invited another three colleagues and we decided to focus on the opportunities for science education to contribute to create peaceful societies as a highly relevant theme for a country undergoing a peace process.

We did not have much funding, but we all had the passion and commitment for social justice. Thus, as our course was not part of any specific academic program and was taken by students and professional alike, then we had the freedom to be as creative as we wanted. From such an open ended course, with the focus on peace, the most important learning I gained was what the students stated as a main priority for Colombia: "Conciliacion" (Conciliation). Conciliacion was described by them as a concept that challenges fundamentalist attitudes which do not take into consideration alternative and confronting perspectives. Conciliacion aims to create dialogue between diverse people with contrasting ideas and above all to forgive and care for each other. For this aim they considered agency, resistance and transformation as priorities of education. What an amazing learning I gained. It was only thanks to this opportunity to create a course out of my passion to contribute to social justice in the place I grew up, without any restrictions in the themes or structure of the course, that I was able to learn about this concept in such a significant context. It is now embedded in my practice and a priority for my teachings back in Australia.

23.3 Life as an International Academic; Is Science as a Site for Resistance a Utopia?

As Derek Hodson argued in 2003, science education could play a role in educating towards resistance, agency and activism. Many of the current social and environmental issues are informed by scientific knowledge and practices. Moreover, as argued by Matthew Weinstein (2016):

> If science has the capacity to tolerate multiple partial truths, rather than a single Truth, then there is a place for the local, for the personal, for the emotional, and for students' own theories and models of nature and life. (p. 243)

However, science education is often presented as a professionalization field, lacking spaces for debates and development of agency. The rearticulation of science into Science Technology, Engineer and Mathematics (STEM) has been embraced internationally. Regarding STEM, some authors such as Weinstein warn us:

> In moving from science to STEM, technical fields are linked into mutually reinforcing and referential sphere, and critically, science is unlinked from issues of history, sociology, and ethics. Engineering plays a particular powerful role in this articulation ... moving from science to STEM is more than just a gathering of like fields and is instead a neoliberal transformation of the field of science ... an appropriation of science to corporatists ends. (2016, p. 238–239)

As I experience international meetings of academics from rich and western countries, I feel there needs to be more of a push towards transformation of the field of science education. For example, during the 2016 NARST meeting, the plenary session speaker, Professor Arnetha F. Ball discussed the theme "Equity, Justice and Generativity in Education Research for Quality Teaching and Learning". Drawing on her Model of Generative Change, Arnetha discussed "practices needed to engage learners across cultures", including how to improve the gap between the academic outcomes of poor and underrepresented communities and those from wealthier groups (Arnetha Ball personal communication April 14th 2016). While listening to her I could not stop questioning what academia and tertiary education could offer to vulnerable or poor communities. Why does disengagement or low outcomes in science education need to be what we focus on? Are we, academics, validating the needs and interests of marginalized groups, or, are we validating the limiting views of what is of value for western countries? Could this discourse that portrays STEM as fundamental to a successful life be another way of colonising other communities? I question whether I should continue participating in international conferences such as this one and how much they could offer me to increase my contribution to social justice.

Similarly, in Australia, we have a national science education conference, ASERA. However, there are no interest groups and no clear opportunities to open spaces for discussing diverse views or consider topics such as Indigenous Knowledge or Science for resistance/action. This was evident when in 2013, with another two colleagues from other Australian universities, we contacted the ASERA organizing committee to ask how we could submit a proposal for a pre-conference workshop related to Indigenous Science. The response was negative, stating ASERA did not support pre-conference workshops despite the fact that we have participated in another such workshop the previous year. Less than a month after this response I received an email from ASERA and colleagues from another university in Victoria, who belong to the ASERA committee, with information of a pre-conference workshop to be offered. My colleague, with whom we were discussing the pre-conference workshops on Indigenous science stated:

> It is my opinion that ASERA serves a very limited research focus and, for this reason, has not extended its participant base over the two decades I have attended. It might be best to simply let it serve this narrow focus and seek out collaborative discussion and action in a different environment. (personal communication 2013)

I sympathised with him and shared his view. Since then I have chosen not to participate in this conference until there are clear spaces open for diversity of opinions and groups. More than providing opportunities to participate in the dominant structures of the west, theories and models from non-dominant groups could inform practices of resistance in education. Pedagogies of resistance originally described in South America, encompass 'reciprocity, solidarity and horizontalidad, or democratic and horizontal decision-making structures' (Bajaj 2015, p. 157). Other theories and approaches, such as Critical Peace Education 'also resist the forces towards regulation, universalization, and the development of rigid norms and standards for what peace education ought to be' (Bajaj 2015, p. 156). These multiple views of education as transformation offer clues to challenge those views of science and environmental education which position ourselves, nature and scientific knowledge as a way of "banking nature" (Freire 1973).

23.4 Identity, Resistance, Opportunity

I consider myself a border-crosser, I identify myself with several communities, approaches and discourses, some of which are distant to academia. I consider the academic profession nowadays can be very limiting and does not tolerate difference and confrontation. The question remains: where can I find opportunities to pursue my view of education as fundamental for social change?

Answering such question is a work in progress; a work shaped daily by all the changes that universities are undergoing and the pressure of a corporate education model. I found the most inspiration not within academia but outside. For instance, I have met over the year different social and environmental activists, whose lives are informed and guided by their passion and commitment towards transformation of the world. For example, I have invited activists from Greenpeace to my classes and worked in international projects with activists who led the movement to ban bullfights in some states in Spain and South American countries. As their cause is connected to social and environmental justice, I consider they could trigger transformation in others.

I wonder: What if environmental, animal protection or social activists who work toward promoting social and ecological justice visit schools or find spaces to meet with tertiary education students who will become school teachers? I have organized programs with this goal in mind. Not only their views enrich my views but also they empower future teachers (Carter et al. 2014). However, creating partnerships with people working in the field, practicing social justice, represents many challenges, particularly in schools when science is pushed to be considered a professionalization discipline, part of the STEM movement.

In Australia, partnerships between teachers and STEM professionals have been a popular program in schools aiming to increase students' engagement and participation in STEM professions. Scientists and Mathematicians in Schools (SMiS), for example, is an Australian program managed by the Commonwealth Scientific and Industrial Research Organisation (CSIRO) on behalf of the nation and funded in collaboration with the Australian government. Established in 2007, by June 2016 SMiS has created partnerships with 2581 schools. Some of the reported outcomes suggest that SMiS program has been successful in increasing students' considerations of STEM careers (STEME Research group 2015), despite the 3.7% unemployment rate of STEM- qualified people (Office of the Chief Scientist 2016). Meanwhile, finding support at the university level or from professional bodies to create partnerships with activists and NGO's have proven to be almost an utopia. I have managed it so far, but only because of the kindness of the activists and their passion to spread their message.

Another program I have created to open opportunities to re-think education towards social justice is to take my 250+ students, pre-service teachers in the final years, to teach diverse approaches to science in marginalized communities in Australia. The demographics of Australian practicing and future school teachers show that the level of multiculturalism presented in the society is not reflected in the teachers' backgrounds. In primary schools, 16.4% of the teachers were born overseas and 19.2% for secondary teachers (McKenzie et al. 2014). Compared to the Australian population that was born overseas at time of the 2006 Census (24%) (Collins and Reid 2012), and 27% by 2011 (McKenzie et al. 2014), immigrants are underrepresented in the teaching sector. Unless teachers are well prepare to educate communities from distinctive cultural and social backgrounds from their own, children from non-English speaking countries will likely see a distance between their needs, interests and what is presented in the classroom. Working towards this goal, during the last 5 years I have led a program in which student teachers have to prepare and teach a unit of work for a particular socio-economic disadvantaged community focusing on the needs and interests of the community.

Despite the interest of some of my students, there is still a high number of students who disengaged from this experience, expressing comments in their class evaluations such as: "The practical experience in the school, could have been just one or two select lessons, as it was a long distance to travel for some of us" or the following comment from one of my master degree students: "While the intention behind the school visits was understandable, removing this one hour session a week and replacing it with a science lab lesson would have been much more beneficial to our learning as teachers." As with the first quote, many students taking my units complain because of the distance to the communities where we carry out the school visits (it takes 45–60 min to get to these schools). Other students, similarly to the second one, ask for learning more common practices in science, such as "lab lessons" with no discussion of the relevance of such experiences. This leaves me with a feeling that my message of the importance of re-thinking science education for social justice often gets lost in the practicalities of the educative practice and the professionalization focus of the discipline which does not sees education as a tool for transformation.

My journey is a journey inspired by my own experience, passion and commitment. But it is not an easy one. It is a journey which often feels utopian and one that requires strength and commitment to resist the pressure and interests of the corporate world. As Paul Thomas (2013) argues:

> Calls for national standards, increased testing, and stringer teacher accountability tied to those standards and tests are in fact efforts to dis-empower and dehumanized students and teachers in ways that feed a corporate/statist machine that sees people as cogs, interchangeable cogs that are valuable only as much as they promote efficiency and profit. (p. 232)

Moreover, as Matthew Weinstein argues, "factors like mandatory, time-consuming testing, and the intensification of an audit culture, limit our ability to resist" (2016, p. 244). My personal journey will hopefully contribute to create new, more creative and critical spaces for resistance and transformation. Similar to Edward Soja's (1996) notion of "thirdspace", my journey aims to create new atypical, non-dominant spaces that will provide the possibility to envision multiple, contrasting and equally valued views and practices for contributing beyond writing papers and delivering lectures. Only then we will be able to transcend limiting, colonizing, hegemonic and universalizing forces.

References

Bajaj, M. (2015). Pedagogies of resistance' and critical peace education praxis. *Journal of Peace Education, 12*(2), 154–166. https://doi.org/10.1080/17400201.2014.991914.

Carter, L., Castano-Rodriguez, C., Jones, M. (2014). Transformative learning in science ducation. In J.L Bencze, & S. Alsop. (Eds.), *Socio-political activism and science & technology education*. Dordrecht.

Collins, J., & Reid, C. (2012). Immigrant teachers in Australia. *Cosmopolitan Civil Societies Journal, 4*(2), 38–61.

Departamento Administrativo Nacional de Estadisica – DANE. (2016). *Comunicado de Prensa: Pobreza Monetaria y Multidimensional en Colombia 2015*. [online] Bogotá: DANE. Available at http://www.dane.gov.co/index.php/estadisticas-por-tema/pobreza-y-condiciones-de-vida/pobreza-y-desigualdad/pobreza-monetaria-y-multidimensional-en-colombia-2015. Accessed 18 Oct 2016.

Ewick, P., & Silbey, S. (2003). Narrating social structure: Stories of resistance to legal authority. *American Journal of Sociology, 108*(6), 1328–1372. https://doi.org/10.1086/378035.

Factor, R., Kawachi, I., & Williams, D. R. (2011). Understanding high-risk behavior among non-dominant minorities: A social resistance framework. *Social Science & Medicine, 73*(9), 1292–1301. https://doi.org/10.1016/j.socscimed.2011.07.027.

Foucault, M. (1984). In P. Rabinow (Ed.), *The foucault reader*. New York: Pantheon Books.

Freire, P. (1973). *Education for critical consciousness*. New York: The Continuum.

Hodson, D. (2003). Time for action: Science education for an alternative future. *International Journal of Science Education, 25*(6), 645–670. https://doi.org/10.1080/09500690305021.

International Displacement Monitoring Centre – IDMC. (2015). *Global overview 2015: People internally displaced by conflict and violence*. [online] Geneva: Norwegian Refugee Council. Available at: http://www.internal-displacement.org/assets/library/Media/201505-Global-Overview-2015/20150506-global-overview-2015-en.pdf. Accessed 18 Oct 2016.

Jaramillo, N., & Carreon, M. (2014). Pedagogies of resistance and solidarity: Towards revolutionary and decolonial praxis. *Interface: A Journal for and about Social Movements, 6*(1), 392–411.

McKenzie, P., Weldon, P., Rowley, G., Murphy, M., McMillan, J. (2014). Staff in Australia's schools 2013: Main report on the Survey 2014. Australian Council for Educational Research. Available at https://www.acer.edu.au/sias. Accessed 22 Oct 2016.

Mezirow, J. (1991). *Transformative dimensions of adult learning*. San Francisco: Jossey-Bass.

Office of the Chief Scientist. (2016). *Australia's STEM workforce: Science, technology, engineering and mathematics*. Camberra: Australian Governments Available at http://www.chiefscientist.gov.au/wp-content/uploads/Australias-STEM-workforce_full-report.pdf. Accessed 22 Oct 2016.

Ogbu, J. U. (2004). Collective identity and the burden of "acting white" in black history, community, and education. *The Urban Review, 36*(1), 1–35. https://doi.org/10.1023/b:urre.0000042734.83194.f6.

Reed-Danahay, D. (Ed.). (1997). *Auto/ethnography: Rewriting the self and the social*. New York: Berg.

Soja, E. (1996). *Thirdspace: Journeys to Los Angeles and other real and imagined places*. Oxford: Blackwell.

STEME research group. (2015). Building productive partnerships for STEM education: evaluating the model & outcomes of the scientists and mathematicians in schools program 2015. Deakin University. Available at http://www.scientistsinschools.edu.au/evaluation.htm. Accessed 22 Oct 2016.

Thomas, P. L. (2013). Corporate education reform and the rise of state schools. *The Journal of Critical Education Policy Studies, 11*(2), 203–238.

Weinstein, M. (2016). Imagining science education through ethnographies of neoliberal resistance. *Mind, Culture, and Activity, 23*(3), 237–246. https://doi.org/10.1080/10749039.2016.1201843.

Worldwide independent network of market research – WIN and Gallup International. (2015). WIN/Gallup international's annual global end of year survey reveals a world of conflicting hopes, happiness and despair. Available at http://www.wingia.com/web/files/richeditor/filemanager/WINGIA_Final_Press_Release_EOY_2015.pdf. Accessed 21 Oct 2016.

Carolina Castano Rodriguez is a lecturer in Science and Environmental Education and the Victoria Coordinator for the National Exceptional Teaching for Disadvantaged Schools (NETDS) Program at Australian Catholic University. She has carried out research projects in Australia, Colombia, Ecuador and Argentina, incorporating transformative learning, ethics of care and critical theories to science education within socioeconomic disadvantaged populations. Her main interest is in the link between science education, social justice and ecojustice.

Chapter 24
Woman Being Disruptive: Challenging (E)quality in Science Education

Annette Gough

I think I was born to be disruptive. I was my parents' first and only child, a redhead and left handed, and I arrived after my parents had been married for nearly 9 years. That I was a girl was not an issue for my parents, or grandparents, but my male cousins saw me as a problem as I was very much doted on by our mutual grandmother and they felt neglected.

Although I grew up in the city, my father loved the Australian bush and being on farms, and he encouraged my interest in animals, the natural environment and science writ large. It seemed taken for granted that I would study science at university, but the end of Year 10 at high school marked a significant point in my personal academic history, for it was then that I had to decide between following an arts (humanities) or sciences stream for my final 2 years of schooling. The decision was not an easy one, as I loved reading and studying history, yet I was also passionately interested in topics relating to the biological sciences. The sciences won, mainly because of the prerequisites needed to get into a science degree at university, and not wanting to close off options at this early stage of my "career". I was fortunate that my school encouraged girls to study science subjects, and I had several female science teachers as role models. My mother was an avid reader, and she helped convince me that I could continue to study history in my own time, or take up history studies later (which I did through undergraduate and postgraduate studies of the history and philosophy of science), and as I have frequently done since, through documenting the history of the field of environmental education – initially as an interpretive case study (Greenall 1981), then from a feminist poststructuralist perspective (Greenall Gough 1994); but also documenting the field (Gough 1997, 2013). Nevertheless, at the time I felt that I was, what I would later call, a "frustrated historian" (Gough 1998b).

A. Gough (✉)
RMIT University, Melbourne, VIC, Australia
e-mail: annette.gough@rmit.edu.au

J. Bazzul, C. Siry (eds.), *Critical Voices in Science Education Research*, Cultural Studies of Science Education 17,
https://doi.org/10.1007/978-3-319-99990-6_24

I pursued undergraduate studies in science education at the University of Melbourne, majoring in botany, zoology and genetics together with education, and became a high school science and biology teacher for a short time. The schools in which I taught were girls' schools, and I experienced almost stereotypical resistance to science in both of them. During my 2nd year I changed jobs and career directions, and relocated to the Curriculum Development Centre (CDC), a potential major change (and disruption) agent for school education in Australia in the 1970s.

Because of the politics involved in working for a national government agency that had to work closely with state governments and non-government organizations (such as teacher organizations) and the subject matter that was my focus, I became intimately aware of the resistances to the fields I have worked in – science and then environmental/sustainability education – and since this time have endeavored to draw attention to those resistances and to disrupt and change practices through curriculum materials development and then my academic teaching and research.

My scholarship and practices are, and have been, stories of disruption, resistance and resilience. In my pursuits I have been consistent with Jacky Colliss Harvey (2015, p. 153, quoting McCracken 1995) who argues, "We cannot rely on them [female redheads] to embrace stereotyped qualities of femaleness – sweetness, docility and politeness… We imagine them ready to give vent to what we keep harnessed." Indeed, this autoethnography has been written as an "'effective' history [which] differs from the history of historians in being without constants", "confirms our existence among countless lost events, without landmark or point of reference" (Foucault 1994, pp. 380–1) and cultivates "the details and accidents that accompany every beginning… scrupulously attentive to their petty malice" (Foucault 1994, p. 373). Michel Foucault (1994, p. 382) also notes, "the final trait of effective history is its affirmation of a perspectival knowledge [*savoir*]". Thus, rather than erasing myself from this work as a traditional historian would do, I have been explicit in tracing the theoretical frameworks I have used, the arguments that I have made, and continue to make, about the oppressions and injustices that science education in schools continues to reproduce due to (e)quality and globalization issues, and what could be done differently to overcome students' continued resistance to science education. I also pay attention to a diverse and specific range of experiences, one's that seem anterior to history but actually tell my story in a rich and wide-ranging way.

24.1 Finding Direction

When I first started teaching high school science in girls' schools in the mid 1970s I was surprised by students' lack of engagement with science in my classrooms. I had always been interested in science and the natural world, and had been encouraged in this by my parents. I thought such interest was normal so to encounter resistance to learning science was unexpected. Yes, some of the things I had had to learn and that I had to teach were boring, but I always felt that there were more good bits than bad – but many of my students disagreed. Was it the science content, my teaching, something else or a combination of these that was not engaging?

I was fortunate that I began my teaching career at the same time as an innovative curriculum project for high school science, the Australian Science Education Project or ASEP, was being developed, and I had the opportunity to pilot some of these materials during my practicum and then, 2 years later, I became involved in their dissemination through the Curriculum Development Centre (CDC). I made this shift from classroom teaching to curriculum work, as I believed that the ASEP materials provided a better way to engage students with learning science. The ASEP materials were very different from the textbook based science that dominated classrooms: there were 41 "relatively independent units so that the teacher has a degree of choice in which ASEP units to use and in what sequence", they catered for "individual differences by providing student options within each ASEP unit and by allowing students to proceed through the materials at their own rates", and they were activity based to engage students in inquiry strategies (Fraser 1978, p. 417).

Given my ecological sciences background, another attractive aspect of ASEP was its underlying philosophy, called the environmental scheme, which, in the sexist language of the time, focused on the "five ways in which man functions: as individuals; as members of interacting groups; with man-made procedures and devices extending his sensory perceptions; in a technological society, the products of which affect man and the natural environment; and in a naturally changing environment" (Lucas 1972, p.446).

Within months of starting to work on ASEP dissemination I also became involved in establishing environmental education as a national priority area for CDC, as documented in my Masters' thesis (reported in Greenall 1981) and my doctoral dissertation (Greenall Gough 1994), thus beginning a lifetime career of working in both science and environmental education.

24.2 Becoming Political

The work at CDC was almost inevitably political because it was a federal agency, so projects involved working with state education authorities and trying to convince them to participate in initiatives. My role was to be disruptive, initiate institutional changes, improve teaching practice, and promote changes to both the curriculum content and teachers' pedagogical practices through new materials and projects – because I believed that the projects I was involved with would lead to better science (and later environmental) education for students, I was as disruptive as I possibly could be. I wrote articles for science teachers' journals (for example, Greenall 1979), ran workshops at teachers' conferences and made presentations to State education department meetings, all with the goal of changing curriculum content and practices.

My first experiences of the political nature of this work were the resistances to supporting the dissemination of the ASEP materials in some states, but political negotiations moved more to the fore when I moved to establishing and then coordinating the CDC environmental education programs, and later when I directed the environmental education section of the Australian Government's environment department (under various names) for several years. These political experiences also informed my critical, and later poststructuralist, predispositions in my academic career. However, as the focus of this chapter is on science education, my environmental education journey takes a back seat, a story for another day, although, as I discuss, environmental education has continued to be interwoven into my science education scholarship over the years. Hence we skip 15 years and take up my story again in 1990 when I started my academic career.

24.3 Becoming Academic

A major focus of my teaching at Deakin University was science education for future primary school teachers. My goal was to make them competent and confident enough to teach science as, at this time, there was very little science education happening in primary schools. A decade later science in the Australian primary school curriculum was still a rare event. Denis Goodrum et al.'s (2001, p. 93) national research study of primary teachers' estimated that the average time spent teaching science each week was 59 minutes. Simon Crook and Rachel Wilson's (2015) more recent study reported a small increase to an average of around 1.6 hours per week across Australia, and this is in spite of significant efforts to encourage the teaching of science in primary schools (as detailed by Crook and Wilson).

At this time there was concern about students' lack of interest in science, but the problem was often seen in simplistic terms, rather than as an interplay of teacher confidence and competence, the science curriculum, and students' understandings of the portrayals of science and their identity construction in science classrooms. For example, Peter Fensham (1985, p. 421) bemoaned the creation of intellectual conceptions of science:

> Science educators, part of the educated in science, have tended to set out to create science education for schools that mirror their own (or science's) priorities. Hence the emphasis on conceptual knowledge and on the intellectual processes that are used with, and in the generation of, this sort of knowledge.

However, while there have been considerable reforms of curriculum for the middle years of schooling in recent years, there is little evidence that these efforts have resulted in improved levels of achievement or attitudes to science (Marginson et al. 2013), despite research, such as that by Rosalind Driver et al. (1996, p.142) which concluded that, "the way science is portrayed in the school science curriculum has a major part to play in shaping students' views of science".

I was concerned about the images of science my teacher education students brought to class, and the images of science my daughter was forming in primary school. These images were consistent with constructivist views of science teaching and learning and other cognitive learning research since the 1980s (for example, Fensham et al. 1994; Tobin et al. 1990) that focused on the disembodied student and cognitive learning. The concern of this research has been with students' mental constructions of science rather than with how they construct their bodily selves and identities within science; the emphasis has been psychological rather than ontological. Driver (1989, p. 85) makes this point quite clearly in her definition of constructivism as "the perspective... whereby individuals through their own mental activity, experience with the environment and social interactions progressively build up and restructure their schemes of the world around them".

I addressed my concerns through several research projects in the early 2000s, and related writings. In one paper associated with a project on improving middle years mathematics and science (Gough 2008a) I looked at how the media constructs stereotyped images of science and scientists (Haynes 1994) and how students perceive science and construct their science identities (or reject having a science identity) in the light of these images. In particular, I argued that students aren't disembodied subjects in our science classrooms as they are influenced by media images of scientists and their reactions to these, and construct themselves accordingly, and concluded that, in science classrooms, we need to

> change the pedagogy and design curriculum which is grounded in students' interests and current issues (such as a science-technology-society-environment orientation) rather than teachers covering what they think needs to be there and teaching it as dry facts and procedures. But much more is needed to interrupt the dominant discourses and engage the discourses that the students bring to science classes. Science teachers need to engage with the media images of scientists with which the students are familiar and students' attitudes to science. (Gough 2008a, p.12)

In a book chapter I also argued that we need to disrupt the dominant discourses by recognizing that gender, equity, equality, quality and globalisation are political issues, which are interwoven into the discourses and practices of science education, and education writ large (Gough 2007). Here I concluded that there is a need for a rethinking of "girls and science" and developing a more democratic science education. Such a democratic science education examines "Western science's complicity with racist, imperialist, [gendered] and Eurocentric projects [and] enables us to gain a more critical, more scientific perspective on an important part of that Western 'unconscious'" (Harding 1993, p. 19).

I also expanded on my arguments to reconceptualize science education and make it more meaningful to students by developing a mutualistic relationship with environmental education in two articles (Gough 2002, 2008b). Given the global decline of student interest in science in schools, and that environmental or sustainability education was often marginalised in the curriculum, I argued that it was to both areas advantage to come together in a major reconstruction of science education. The theoretical arguments in these articles were supported by stories from a research project which exemplified how primary schools students were interested in the environment and chose environment related topics for their science projects when given a choice (Gough and Sharpley 2005).

24.4 Developing Gendered Postcolonial Perspectives

The feminist studies of science, and science education, that were gaining traction in the late 1980s and early 1990s caught my attention at this time, both for my own doctoral research purposes (Greenall Gough 1994), but also as a critique of masculinist science education (Gough 1998a). Jane Butler Kahle's work (1987, 1988, 1990), for example, focused on science stereotypes and "Draw–A–Scientist Test"-ing (DAST) as a means of measuring the effectiveness of intervention programs towards a more equitable science education.

Because I was teaching science education through the use of popular culture (children's literature, movies, television programs, newspapers and magazines) as much as possible to make the students feel comfortable, I was aware of R. A. Schibeci's (1986, p.146) argument that,

> Images of science and scientists in popular culture represent a proportion (possibly not a very significant proportion) of the range of practices and behavior in modern science. There is sufficient evidence to indicate that scientists engage in a very diverse range of behavior, of which the more outlandish have been seized on by those who control the images presented in popular culture.

Also, the discourses available to students with respect to the images of science and scientists are perhaps limited – as indicated by the stereotypes described by Roslynn Haynes (1994). Thus in my teaching I tried to make science as familiar as possible by showing how it could be taught using children's books such as Pamela Allen's (1980) *Mr Archimedes' Bath* and movies such as *Jurassic Park* (Kennedy et al. 1993).

As my research and writing evolved during the 1990s and into the 2000s I became increasingly concerned with how images and metaphors have a constitutive force in identity formation. I argued that it is important for students that they deconstruct the images in popular culture and elsewhere in their experiences, and recognise that there are wider options available for them to consider in terms of gender, colour, sexuality, ethnicity, class, ablebodiedness and other aspects of signification in asserting their identity in science classrooms. Such notions of identity and constructions of self in science classrooms were relatively new to science education research, and some science educators still probably consider such notions to be heretical.

In the mid 1990s Jane Kenway and I were asked to review gender and science education in schools, which we did "with attitude" (Kenway and Gough 1998). Here we agreed that "gender equity research ought to transcend the boundaries of race, ethnicity, class and socio-economic identities" (Krockover and Shephardson 1995, p. 223), and this increasingly became my research focus, in a search for "pedagogies of science (in)formed by global perspectives" (Gough 2001). In this chapter I focused on the apparent lack of awareness/consciousness among science educators (as well as scientists) of the global culture of which science is a part, and discussed the need to change both the knowledge and classroom practices of science educators, particularly through adopting positions such as Sandra Harding's (1993, 1998) 'strong objectivity.' Rather than throwing out the baby with the bath water and totally reject Western science, Harding recognizes "some of the still immensely valuable elements of the European philosophic tradition' and argues for retrieving and transforming them 'into tools useful in today's multicultural and postcolonial world" (1998, p. x). Both Harding (1993, 1994, 1998) and Carolyn Merchant (1980) led me to believe that we need to recognize the socially constructed, gendered and multicultural nature of science in its global context within science education pedagogies and to work towards a more democratic concept of science in our science teaching practices, and encourage the development of postcolonial and poststructuralist student identities in science classrooms. By doing this teacher's should have a greater likelihood of engaging students in learning science; a science education more connected with societal issues and their lives.

24.5 Staying Disruptive

The shift to a multicultural science education and the challenging of the paradigm of what science is, and transforming the science curriculum is largely still to happen – especially taking into account the race, class, ethnicity and socio-economic aspects – and I continue to feel frustrated. As a result of my trans-Indian Ocean experiences I argued that

> There is a need for some fundamental changes in the way science is represented in schools – moving away from science as objective and dispassionate, reassessing the nature of evidence and explanation (and their relationship to each other), and reviewing the status of scientific knowledge, especially the philosophical and psychosocial aspects of learning environments. (Gough 2007, p. 144)

Such an approach still overlooks students' identity construction in science classrooms and, whatever identity the students are constructing for themselves within science the overall conclusion is the same: although the students strongly agree that science and technology are important for society (Sjøberg and Schreiner 2005), most do not see themselves as part of it and they are not interested in pursuing a future career in science or interested in studying science at secondary school (Goodrum et al. 2012).

Students' images of scientists are not being engaged in their science classrooms, yet being aware of the images that students hold of science and scientists is an important consideration in engaging them with science: "mental imagery is one of the most important of all human abilities: it enables many everyday tasks, such as navigation and understanding of verbal descriptions... and helps us to perform high-level activities such as creating art and doing research" (Howard 1992, p. 33). The area of science education research concerned with personal constructivism takes some of these aspects into account, albeit from a cognitivist perspective. However, much of this research treats science and scientists as objects for study by students. They are engaged at the level of the mind, but not the subjective body. Although treated as if disembodied, students construct themselves in science classrooms in a number of ways. They bring with them a range of previous experiences and media images, as well as understandings, beliefs, attitudes and behaviours. The teacher, the texts, the discourses and their fellow students position them. They construct themselves and perform their identities within this context. However, the question remains, how can we get teachers to engage with students' identity construction in science classrooms?

It has become obvious in recent times that the STEM movement and the standardizing of curriculum is not the answer. While various governments' instrumentalist STEM agenda include lifting the overall scientific literacy of the population and attracting more students to study STEM at secondary and university levels, they see this as happening by changing the students, not changing the science (education). Science education remains in crisis. Numerous government documents generally acknowledge that there is a declining interest in studying such areas from school and prospective university students (Goodrum et al. 2012), and declining public scientific literacy (Wyatt and Stolper 2013). As I argued recently (Gough 2015), there are many aspects of this discussion that remind me of the movie *Field of Dreams* (Gordon et al. 1989)— the reports believe that if the universities build the programs then the students will come and study them—but will they, and who will teach them and what will they be taught? The lack of recognition of these issues is very frustrating, but it is important to continue to be disruptive.

Most recently my scholarship frequently focuses on critiquing the Australian science curriculum for being instrumentalist and disconnected from sociopolitical issues, and for the teaching workforce not being prepared to teach such issues (Gough 2015). I have also critiqued it for being

> an example of what Foucault (1978) calls bio-power, which is not just governmental power but a power that is invisible, plural, discursive, pervasive and enforced, via a plethora of power relationships, to manage human life, bodies and species for economic and political ends. (Gough 2017, p. 891)

As long as science educators and science education curricula continue to ignore engaging students' identity formation in science classrooms and making the curriculum more connected to socio-political issues and students' lives I will continue to be disruptive. I have a passion for science, but not as it is currently taught in schools and I want to see this changed.

References

Allen, P. (1980). *Mr Archimedes' bath*. Sydney: Angus & Robertson.

Crook, S., & Wilson, R. (2015). Five challenges for science in Australian primary schools. *The Conversation*. Retrieved on 1 Dec 2016 from https://theconversation.com/five-challenges-for-science-in-australian-primary-schools-42413.

Driver, R. (1989). The construction of scientific knowledge in school classrooms. In R. Millar (Ed.), *Doing science: Images of science in science education* (pp. 126–136). London: Falmer Press.

Driver, R., Leach, J., Millar, R., & Scott, P. (1996). *Young people's images of science*. Buckingham: Open University Press.

Fensham, P. J. (1985). Science for all. *Journal of Curriculum Studies, 17*(4), 415–435. https://doi.org/10.1080/0022027850170407.

Fensham, P., Gunstone, R., & White, R. (1994). *The content of science: A constructivist approach to its teaching and learning*. London: Falmer Press.

Foucault, M. (1978). *The history of sexuality, volume one, an introduction* (Robert Hurley, Trans.). New York: Pantheon.

Foucault, M. (1994). In J. D. Faubion (Ed.), *Aesthetics, method, and epistemology* (trans: Hurley, R., & others). New York: The New Press.

Fraser, B. J. (1978). Australian science education project: Overview of evaluation studies. *Science Education, 62*(3), 417–426. https://doi.org/10.1002/sce.3730620319.

Goodrum, D., Druhan, A., & Abbs, J. (2012). *The status and quality of year 11 and 12 science in Australian schools, A report prepared for the Office of the Chief Scientist*. Canberra: Australian Academy of Science.

Goodrum, D., Hackling, M., & Rennie, L. (2001). *The status and quality of teaching and learning of science in Australian schools*. Canberra: Department of Education, Training and Youth Affairs.

Gordon, L., Gordon, C. (Producer), Robinson, P.A. (Director). (1989). *Field of dreams* [motion picture]. Universal Pictures.

Gough, A. (1997). Founders of environmental education: Narratives of the Australian EE movement. *Environmental Education Research, 3*(1), 43–57. https://doi.org/10.1080/1350462970030104.

Gough, A. (1998a). Beyond eurocentrism in science education: Promises and problematics from a feminist poststructuralist perspective. In W. F. Pinar (Ed.), *Curriculum: Toward new identities* (pp. 185–209). New York: Garland.

Gough, A. (1998b). (W)riting curriculum histories. In J. Mousley, N. Gough, M. Robson, & D. Colquhoun (Eds.), *Horizons, images and experiences: The research story collection* (pp. 109–117). Geelong: Deakin University.

Gough, A. (2001). Pedagogies of science (in)formed by global perspectives: Encouraging strong objectivity in classrooms. In J. A. Weaver, M. Morris, & P. Appelbaum (Eds.), *(Post) modern science (education): Frustrations, propositions, and alternative paths* (pp. 275–300). New York: Peter Lang.

Gough, A. (2002). Mutualism: A different agenda for science and environmental education. *International Journal of Science Education, 24*(11), 1201–1215. https://doi.org/10.1080/09500690210136611.

Gough, A. (2007). The politics and practices of equity, (e)quality and globalisation in science education: Experiences from both sides of the Indian Ocean. In B. Atweh, A. C. Barton, M. C. Borba, N. Gough, C. Keitel, C. Vistro-Yu, & R. Vithal (Eds.), *Internationalisation and globalisation in mathematics and science education* (pp. 129–147). Dordrecht: Springer. https://doi.org/10.1007/978-1-4020-5908-7_8.

Gough, A. (2008a). Addressing the science in schools crisis: media images, scientific literacy and students' lives. Refereed papers from the 2007 Annual Conference of the Australian Association for Research in Education, Fremantle, W.A. Retrieved 4 Jan 2017 from http://www.aare.edu.au/publications-database.php/5372/Addressing-the-science-in-schools-crisis:-Media-images,-scientific-literacy-and-students'-lives.

Gough, A. (2008b). Towards more effective learning for sustainability: Reconceptualising science education. *Transnational Curriculum Inquiry, 5*(1), 32–50 Retrieved 4 Jan 2017 from http://nitinat.library.ubc.ca/ojs/index.php/tci/article/viewFile/90/269.

Gough, A. (2013). The emergence of environmental education: A 'history' of the field. In R. B. Stevenson, M. Brody, J. Dillon, & A. Wals (Eds.), *International handbook of research on environmental education* (pp. 13–22). New York: Routledge for the American Educational Research Association. https://doi.org/10.4324/9780203813331.ch2.

Gough, A. (2015). STEM policy and science education: Scientistic curriculum and sociopolitical silences. *Cultural Studies of Science Education, 10*(2), 445–458. https://doi.org/10.1007/s11422-014-9590-3.

Gough, A. (2017). Searching for a crack to let environment light in: Ecological biopolitics and education for sustainable development discourses. *Cultural Studies of Science Education, 12*(4), 889–905. https://doi.org/10.1007/s11422-017-9839-8.

Gough, A., & Sharpley, B. (2005). Toward effective teaching and learning – stories of primary schools' environmental science interest and action. *Educational Action Research, 13*(2), 191–211. https://doi.org/10.1080/09650790500200284.

Greenall, A. (1979). Innovations in science education – CDC's environmental education project. *Australian Science Teachers Journal, 25*(1), 41–46.

Greenall, A. (1981). Environmental education – a case study in national curriculum action. *Environmental Education and Information, 1*(4), 285–294.

Greenall Gough, A. (1994). Fathoming the fathers in environmental education: A feminist post-structuralist analysis. Unpublished doctoral dissertation. Geelong: Deakin University.

Harding, S. (Ed.). (1993). *The "Racial" economy of science: Toward a democratic future.* Bloomington: Indiana University Press.

Harding, S. (1994). Is science multicultural? Challenges, resources, opportunities, uncertainties. *Configurations, 2*(2), 301–330. https://doi.org/10.1353/con.1994.0019.

Harding, S. (1998). *Is science multicultural?* Bloomington: Indiana University Press.

Harvey, J. C. (2015). *Red: A history of the redhead.* New York: Black Dog & Leventhal Publishers.

Haynes, R. D. (1994). *From Faust to strangelove: Representations of the scientist in Western literature.* Baltimore: Johns Hopkins University Press.

Howard, R. (1992). I spy with my mind's eye... *The Bulletin*, p.33.

Kahle, J. B. (1987). Images of science: The physicist and the cowboy. In B. J. Fraser & G. J. Giddings (Eds.), *Gender issues in science education* (pp. 1–11). Perth: Curtin University of Technology.

Kahle, J. B. (1988). Gender and science education II. In P. J. Fensham (Ed.), *Development and dilemmas in science education* (pp. 249–265). London: Falmer Press.

Kahle, J. B. (1990). Real students take chemistry and physics: Gender issues. In K. Tobin, J. B. Kahle, & B. J. Fraser (Eds.), *Windows into science classrooms: Problems associated with higher-level cognitive learning* (pp. 92–134). London: Falmer Press.

Kennedy, K., Molen, G. R. (Producer), Spielberg, S. (Director). (1993). Jurassic Park (motion picture). Universal Pictures.

Kenway, J., & Gough, A. (1998). Gender and science education in schools: A review 'with attitude'. *Studies in Science Education, 31*, 1–29. https://doi.org/10.1080/03057269808560110.

Krockover, G. H., & Shepardson, D. P. (1995). Editorial: The missing links in gender equity research. *Journal of Research in Science Teaching, 32*(3), 223–224. https://doi.org/10.1002/tea.3660320303.

Lucas, A. M. (1972). ASEP – a national curriculum development project in Australia. *Science Education, 56*(4), 443–451. https://doi.org/10.1002/sce.3730560403.

Marginson, S., Tytler, R., Freeman, B., & Roberts, K. (2013). *STEM: Country comparisons.* Melbourne: Australian Council of Learned Academies.

McCracken, G. (1995). *Big hair: A journey into the transformation of self.* London: Indigo.

Merchant, C. (1980). *The death of nature: Women, ecology and the scientific revolution.* New York: Harper and Row.

Schibeci, R. A. (1986). Images of science and scientists and science education. *Science Education, 70*(2), 139–149. https://doi.org/10.1002/sce.3730700208.

Sjøberg, S., & Schreiner, C. (2005). *Young people and science: Attitudes, values and priorities.* Evidence form the ROSE project. Presentation at the EU Science and Society Forum, Brussels, 8–11 March.

Tobin, K., Kahle, J. B., & Fraser, B. J. (Eds.). (1990). *Windows into science classrooms problems associated with higher-level cognitive learning.* London: Falmer Press.

Wyatt, N., & Stolper, D. (2013). *Science literacy in Australia.* Canberra: Australian Academy of Science.

Annette Gough is Emeritus Professor of Science and Environmental Education in the School of Education at RMIT University, Melbourne, Australia. Her research interests span environmental, sustainability and science education, research methodologies, post-human and gender studies and she has completed research projects for national and state governments as well as working with UNESCO, UNEP and UNESCO-UNEVOC on several research and development projects.

Chapter 25
After-Words: Refashioning Science/ Education Through Critical Voices and Politics

Matthew Weinstein

For me this collection of thoughts, reflections, and refigurings points to the problems of our urgent work. Voices, disembodied, relocated in text, are proxies for complex material struggles in dozens of sites working to shape more just, more inclusive, more ethical, and more playful spaces for our human and non-human companions. My own voice, not as meta, but in conversation with these chapters–and people who I know and intersect within a variety of conversations–merely can point to what I see as lacunae in our struggles to move forward in a too often bloody world.

One thing that inspires me in this collection is the return to a reflexivity—through autoethnography, dialogue, etc.—that in no way reproduces Donna Haraway's anxiety that reflection is just displacing the same elsewhere (1997). We have learned, as Haraway's early work urged, that all reflections, all visions, all voices are particular techno-natural fabrications that can be faithfully disassembled, recombined, and otherwise analyzed (1991b). These reflections, in other words, are technologies for accomplishing certain kinds of justice-oriented tasks; they are pragmatic. The challenge is not the classic goal of true representation, but of selecting parts and portions for accomplishing the work of bettering this world.

And I am doing that here. My machinery for bettering a just science world is deeply indebted to Haraway, who first modeled for me a language of critical science discourse. It also weaves together Bruno Latour's problematic figuration of science as network (Latour 1987), Evelyn Fox Keller's psychoanalytic and then realist historical narratives of science as masculine mattering (1985, 1992), Maria Lugones's world-traveling as practice of critical playfulness (1990), Judith Butler's postmodern positing of mattering as political/gendered figuration (1993), Philippe Pignarre and Isabelle Stenger's brewing of Marxism, antiracism, and feminism to envision a post

M. Weinstein (✉)
School of education, University of Washington-Tacoma, Tacoma, WA, USA
e-mail: mattheww@uw.edu

J. Bazzul, C. Siry (eds.), *Critical Voices in Science Education Research*,
Cultural Studies of Science Education 17,
https://doi.org/10.1007/978-3-319-99990-6_25

autre-mondialization politics (Pignarre et al. 2011), Kim Fortun's ethnographic tracing of the toxic subject as the political subject of the ever morphing neoliberal, colonial zeitgeist (1996), and (not even finally), to Sylvia Wynter's insistence that our project of being human is painfully incomplete (McKittrick 2015b). I conjure their spirits to locate myself and my projects in a certain problematic history that makes certain things critical and not others. My sense of injustice is born of male privilege, of Jewish white uncertainty (or rather the certainty that whiteness is a power that is granted by others and can disappear in a flash), of a certain peripatetic life, of one who has lived for long periods on the edge of queer communities but whose pleasures are decidedly norm-approved, of having organized Black and Latinx neighborhoods of early neoliberal Dallas, TX. This is a situated knowledge and my sense of imbalance and balance is always the product of a specific habitus that can only find resonance with some and not others, as my aggravated students will gladly attest.

So one of my troubles has always been the tension between micro and macro struggles. As a teacher I'm lured by the micro: micro analysis, micro politics, and even micro sizing of chemical experiments. There is a secondary teacher in me who wants to close the door, put posters over the panopticon window to keep administrative eyes from prying, and then go deep into conversation with these particular students, in this place, at this time. But, as an ex-organizer, as someone trained by political-economists, as the son of an economist, I am painfully aware that what seems micro, what seems like personal choice, is often determined in ways I cannot imagine by the rules of political, gender, and racial economy. I am ever convinced, even if my secondary teacher mind desires to ignore it, that struggles in the classroom have to parallel struggles outside the classroom, and that they are entangled. To imagine democratic classrooms is to also engage in a struggle for schools to be embedded in community struggles, for parents to work with teachers in ways that teachers will often find problematic, to name and fight those agents who often appropriate equity discourse to mine schools and education writ-large for profit.

And while I am discussing struggles of different scales, I would remind those that read this that science, like schooling, is the imbricated locus of reconception. As that long list of authors earlier cited have all combined to say, science is the vehicle and repository of modeling that was central to maneuvers of gender, national, racial, and colonial power. It's no coincidence that physics both in its basic and applied versions is largely funded out of military and Department of Energy (which acts as an extension of military policy through its nuclear responsibilities) in the U.S. or that the U.S. Public Health Service wears what is basically Naval uniforms. Wynter traces the ways that our current economy of race and money is deeply tied to narratives that fuse Darwinian tales of survival of the fittest and (neo)classical economics. Science is central to the logic of racial and gender economies. Perhaps, nothing better indicates this than the recent March for Science, which sought to protect funding for often critical-friendly projects of measuring and countering global warming, or studying and treating urgent medical priorities. As Jean Aguilar-Valdez (2017) brilliantly analyzes, the language of the March itself produced a language of exclusion and selection: the smarts (or as the marchers symbolically emphasized, "the brains") vs. the stupids. The March thus did not align with struggles for justice, but rather emphasized a conservative (i.e., protective of the scientists' own political authority) platform.

Furthermore, schools, as the machinery of cultural reproduction, as the site where the sciences reproduce themselves, as the site where science-as-modernity is embedded everywhere from the gym to the front office and attendance register, is central to this technoscientific economy. Failure is built in at all levels; legitimation of selection drives operations: schools are, as many have noted, sorting and selecting machines. To use Henry Giroux's dire language, machines for producing "disposable" (or at least repurposable) populations (2009). I have always winked at this blurring of boundaries as to where science education and schooling writ large begin and end through the literary device of the slash, i.e., the way fan fiction, i.e., the stories written and shared by fans of TV, movie, and even books to extend and explore those fictive universes, indicates homoerotic relationships. Fanfiction writers indicate at the start of the story that Bob and Eduardo will be romantically entangled by stating Bob/Eduardo. I mark the entangled, same genre relationship of modern schooling and modern science through the signifier Science/Education.

As a U.S. citizen, the need for more just worlds, including science/educations, is feeling more urgent now. While the world stumbles forward seemingly as before, for, as everyone is quick to point out, the previous administration also engaged in a biopolitics of deporting immigrants, there is an intensification of the discourse, an open racism, a dismissal of knowledge, an embrace of terror and war that I have not known in my lifetime in this country. Gone are the apologetics, the humanist rationales for intervention, and the embarrassment at the racial foundations of the nation-state. I know that we are not alone in being subject to this (alt)rightward slide. Brexit, neo-nationalism in Australia, the rise of extreme xenophobic, homophobic, and sexist orders from the England (Brexit) to Russia, Poland, and the Middle East, are part of co-developed discourse of a conservative, autocratic racial caste. In thinking about refashioning schools, the economy, culture, we cannot ignore the success that these dark forces have had, and that our visions of justice, politics (in the sense of action), and possibility are not determined by, but must be nuanced in relation to this current state of emergency.

So to add my voice to moving forward with the collective and entangled work of the authors here, I want to push for developing languages and politics that can encompass multiple venues and levels of policy and practice without ever simply collapsing micro and meso into macro or vice versa. A science education politics that pushes back against extant science, schooling, economics, and state-craft in parallel and simultaneously. This is a science education politics that frames economics not in some sort of opposition to discourses of race, gender, and sex, but that helps frame gender, sex, and race as produced through sets of exchange and connections, even if those connections appear to be autonomous actions (e.g., personal choices). I was moved that so many contributors frame their work through ethics, which I read as a philosophical rewrite of new economies, for ethics reworks the conditions of obligation and exchange.

So one aspect of these analyses and politics is to join those already engaged in this work. Here, in the U.S. teachers are ahead of us academics in many regards. From the BATs (Badassed Teachers Association; https://www.badassteacher.org/)

who organize to resist neoliberal school reform, to Rethinking Schools (https://www.rethinkingschools.org/), who provide counter narrative lesson plans to the dominant, conservative educational orthodoxy, we must figure out how to help and support the already in motion resistance to the educational models of the grotesquely powerful. The same goes to struggles over political economy at more sweeping scales. My examples come, again, from the U.S., but we have been slow to realize and find a way to fight the enormous power and money behind "education reform." Students and teachers in Brazil, Spain, Chile, and elsewhere are more organized and been more resistant—yearlong strikes in some cases—against (not coincidentally) parallel reform efforts.

But I think there is additional work that our community needs to engage in. Following again Haraway's lead (2016), we need speculative fabulations of how schools and their economies might operate other than as mechanisms of selection and sorting. This is sort of fiction as planning; and is certainly not novel. My urging is that the school again never be framed as autonomous of the communities and economies it serves.

Similarly, science itself needs to be refashioned to fit such hopefully possible worlds. Harding calls such a science a "successor science" (1991). Walter Mignolo following Wynter calls it *scientia* (2015). Many drawing on Wynter, who herself builds on Franz Fanon, position such a next-science on the disciplinary boundaries of the arts, humanities, and sciences (McKittrick 2015a; Mignolo 2015). To decolonize science, science needs to learn a certain polyvocality that it lacks and that the critical humanities have tried to model, e.g., in the fluid code switching of the philosophy of Maria Lugones (1994) and critique of Gloria Anzaldúa (1990). But we should also acknowledge the counter narratives that exist within the sciences themselves, as so much of Haraway's early work points to. In addition to the Darwinian counter narrative to the usual one of survival of the fittest by Petr Kropotkin in his work on mutual aid (1989), there are the tales of immune systems depending on toxins to learn their work (Haraway 1991a); of megafauna as bacteriological codependent colonies (Margulis and Sagan 1995), and the many animal stories that Haraway likes to tell (1992). There is Barad's (2007) entangled ontology, which is, I feel, pushing us to a language of "conviviality" rather than projectile motion and falling feathers within physical science thinking (Illich 1973). These are exemplars and resources for a sort of fan fiction of our own. If capitalist science/education does not satisfy, yet we are drawn to classrooms and schools as sites of hope and care, then we might take a hand at rewriting the story to produce the relationships that we, our students, and their students (for those of us in teacher education) will find more satisfying.

References

Aguilar-Valdez, J. (2017). *Dismantling the white male lab coat, science-as-property narrative: When we define science as human in dehumanizing ways*. Paper presented at the National Association of Research on Science Teaching, San Antonio, TX.

Anzaldúa, G. (1990). *Making face, making soul*. San Francisco: Aunt Lute Foundation Books.

Barad, K. M. (2007). *Meeting the universe halfway: Quantum physics and the entanglement of matter and meaning*. Durham: Duke University Press.

Butler, J. (1993). *Bodies that matter: On the discursive limits of "sex"*. New York: Routledge.

Fortun, K. (1996). *Environment, epidemiology, and the work of culture: Alternative strategies*. Paper presented at the meeting of the American Anthropological Association, San Francisco.

Giroux, H. A. (2009). *Youth in a suspect society: Democracy or disposability?* (1st ed.). New York: Palgrave Macmillan.

Haraway, D. J. (1991a). The biopolitics of postmodern bodies: Constitutions of self in immune system discourse. In D. J. Haraway (Ed.), *Simians, cyborgs, and women: The reinvention of nature* (pp. 203–230). New York: Routledge.

Haraway, D. J. (1991b). Situated knowledges: The science question in feminism and the privilege of partial perspective. In D. J. Haraway (Ed.), *Simians, cyborgs, and women: The reinvention of nature* (pp. 183–201). New York: Routledge.

Haraway, D. (1992). Otherworldly conversations; terran topics; local terms. *Science as Culture, 3*(14), 65–98.

Haraway, D. J. (1997). *Modest_Witness@Second_Millennium.FemaleMan©_Meets_OncoMouse™*. New York: Routledge.

Haraway, D. J. (2016). *Staying with the trouble: Making kin in the Chthulucene*. Durham: Duke University Press.

Harding, S. (1991). *Whose science? Whose knowledge? Thinking from women's lives*. Ithaca: Cornell University Press.

Illich, I. (1973). *Tools for conviviality*. New York: Harper & Row.

Keller, E. F. (1985). *Reflections on gender and science*. New Haven: Yale.

Keller, E. F. (1992). *Secrets of life, secrets of death: Essays on language, gender and science*. New York: Routledge.

Kropotkin, P. A. (1989). *Mutual aid: A factor of evolution* (1st ed.). Montreal: Black Rose Books.

Latour, B. (1987). *Science in action*. Cambridge, MA: Harvard University Press.

Lugones, M. (1990). Playfulness, "world"-traveling, and loving perception. In G. Anzaldúa (Ed.), *Making face, making soul, haciendo caras: Creative and critical perspectives by women of color* (pp. 390–402). San Francisco: Aunt Lute Foundation Books.

Lugones, M. (1994). Purity, impurity, and separation. *Signs, 19* (Winter), 458–479.

Margulis, L., & Sagan, D. (1995). *What is life?* New York: Simon & Schuster.

McKittrick, K. (2015a). Axis, bold as love: On Sylvia Wynter, Jimi Hendrix, and the promise of science. In K. McKittrick (Ed.), *Sylvia Wynter: On being human as praxis* (p. xiii., 290 pages). Durham: Duke University Press.

McKittrick, K. (2015b). Sylvia Wynter: On being human as praxis. Retrieved from https://doi.org/10.1215/9780822375852.

Mignolo, W. D. (2015). Sylvia Wynter: What does it mean to be human? In K. McKittrick (Ed.), *Sylvia Wynter: On being human as praxis* (p. xiii., 290 pages). Durham: Duke University Press.

Pignarre, P., Stengers, I., & Goffey, A. (2011). *Capitalist sorcery: Breaking the spell*. Basingstoke/New York: Palgrave Macmillan.

Matthew Weinstein is a Professor of Science Education at the University of Washington—Tacoma. He is the author of the books *Robot World* and *Bodies out of Control*, the latter with Nidaa Makki and the TAMS Analyzer qualitative research tool. In addition, he has published many articles on social justice, social movements, and public cultures of science.

MIX
Papier aus verantwortungsvollen Quellen
Paper from responsible sources
FSC® C105338

If you have any concerns about our products,
you can contact us on
ProductSafety@springernature.com

In case Publisher is established outside the EU,
the EU authorized representative is:
Springer Nature Customer Service Center GmbH
Europaplatz 3, 69115 Heidelberg, Germany

Printed by Libri Plureos GmbH
in Hamburg, Germany